INTERDISCIPLINARY MATHEMATICS

VOLUME 25

GEOMETRIC COMPUTING SCIENCE: FIRST STEPS

Robert   Hermann

MATH SCI PRESS
53 JORDAN ROAD
BROOKLINE, MA. 02146

Article by R. Hermann reprinted from Acta Applicanda Mathematica,
published by Kluwer Academic Publishers, Dordrecht, Netherlands.

ISBN 0-915692-41-4

Library of Congress Cataloging-in-Publication Data

Hermann, Robert.
    Geometric computing science : first steps / Robert Hermann.
        p.   cm. -- (Interdisciplinary mathematics ; v. 25)
    Includes bibliographical references and index.
    ISBN 0-915692-41-4
    1. Computer science--Mathematics.   I. Title.   II. Series:
Hermann, Robert.   Interdisciplinary mathematics ; v. 25.
QA76.9.M35H47 1991
004'.01'51--dc20   91-13531
                                                                    CIP

GEOMETRIC COMPUTING SCIENCE: FIRST STEPS

PREFACE

This book outlines a mathematical way of thinking about the interaction of 'pure' and 'applied' computer science. I will be more subjective than is customary in science and mathematics exposition and sketch my personal vision, in terms of my own experience in pure and applied mathematics and my extensive readings in the scholarly computer science literature. Unfortunately, I have not been able to bring the ideas I sketch here to the state of completion that I might wish. I am offering them in this incomplete form because I believe that they will be useful to mathematically-minded scientists and engineers who want to understand how the computing science insights fit into the overall mathematical structure of their discipline and to mathematicians who want to think about computing science with the same intellectual tools that have been utilized to great mutual benefit in the mathematization of traditional, differential-equations based science and engineering.

I have been a research mathematician for thirty-five years. In that period, there has been a magnificent flowering of geometric mathematics, and its application in such disciplines as physics, mechanics, control theory, and logic. I believe that the natural progression will be towards the use of geometric methods and ways of thinking as a unifying mathematical methodology in computer science. However, to see it happen in my lifetime will require some push from a group of researchers who are intellectually broad enough to be reasonably familiar with the literature and methodology of both computer science and mathematics. Another reason that I put this book out in this admittedly incomplete and fragmentary form is to provide a challenge to pursue this goal!

The first obvious questions are: What is Computer Science?
What is Geometric Mathematics? What will it mean to combine
the two? I can not give complete answers here, only
fragmentary observations based on my own understanding and
experience.

First, any serious attempt to combine the two disciplines
will find that the algebra of Category Theory will be an essential
intellectual tool, in much the way that the algebra of
polynomials played a role in Descarte's construction of Analytic
Geometry or that Commutative and Grassmann algebra plays a
role in Differential Geometry. Using this crutch, I can pass the
buck down the line and say that by 'Geometric Mathematics' I
will mean the study of the categories and functors that have
been developed to study geometric structures: Manifolds,
topological spaces, cell complexes, coset spaces of groups, etc.;
and geometrically- defined subsets of these spaces - The
'Erlanger Programm', and All That.

My task would be easier if a contemporary Felix Klein of
Computer Science had laid out his unifying vision. However, it
seems that we are still at an early historical stage and the right
concepts have not sufficiently gelled. Much of what is most
vital and important in contemporary computer science is closely
tied to the dramatic engineering advances of the past forty
years, and has not yet been codified in a form that is amenable
to broad mathematical treatment. I want to stand aside from
the 'pragmatic' goals of most of the other computer science
researchers and study the sort of questions that 'fundamental'
physicists ask themselves, inspired by 'experiment', but also
following their own path, guided by their own knowledge and
intuition.

Certainly, I see a central question: What is the
Mathematical Nature of a Computer Program. Perhaps this is
analogous to the question - What is an Elementary Particle? -

that has guided (and still eludes!) physicists for the past forty years. I started out five years ago on the 'pragmatic' side, reading the literature in Artificial Intelligence to see what might be helpful in a control- oriented research program at NASA for which I was doing research. I found it difficult to think at such an 'applied' level without a better understanding of how the fundamental parts of Computer Science fit into the differential-equations oriented world of the physical scientist and engineer. Such concerns have led to this work.

A key mathematical question will be: What is the right categorical notion of 'data'? There are three subsidiary questions: How is that 'data' represented in the machines we call 'computers'? In the 'outside, world'? and how are they related? The attraction for Artificial Intelligence among Computer Scientists is, perhaps, that our own mind provides a very powerful model and example for the development of such a theory. However, it seems to me that jumping off from such an intellectually ambitious position has not been a success: Evolution (and mathematics) goes from the simple to the complex!

Another useful and broad insight has been that the process - first codified in Logic - of defining a Language, i.e. a 'syntax', and then finding 'models', i.e. a 'semantics', is an essential feature of the ultimate Erlanger Programm of Computer Science. Here, certainly, the motivation for the utilization of 'category' and 'functor' concepts becomes evident to a mathematician. Indeed, the books and research papers on Computer Semantics already extensively use the ideas developed in the branch of Algebra called Category Theory. However, I believe that the right algebraic formalism for expressing my ideas will be some combination of the now-standard category story and more concrete algebra (and topology) that reigns in contemporary geometric mathematics.

All of my books (and many of my papers) combine 'exposition' and 'research'. In this book, the 'exposition' is oriented towards the mathematician, scientist or engineer who - without extensive knowledge of Category Theory - wants to understand how it is related to the more conventional mathematics utilized in the physical sciences. Given my own education in the geometric mathematics of the 1950's, I am aware that Category Theory has its roots in ideas current in that period, hence tend to emphasize its connection to that sort of mathematics.

Part of this work was done while I was supported by grants from NASA-Ames and the NSF Applied Math program.

CONTENTS

VII. RECURSIVE SYSTEMS AND JET BUNDLES

VIII. THE EHRESMANN JET-SPACES AND THE PROLONGATION CALCULUS

RECURSIVE SYSTEMS AND GEOMETRIC COMPUTING SCIENCE

ABSTRACT

Ideas from category and sheaf theory, numerical analysis, Lie theory, control theory and differential system theory are stirred together . The interdisciplinary goal is to develop a mathematical framework broad and powerful enough to span the gulf between the 'discrete' world of the computer scientist and logician and the 'continuous', differential-equation based world of the physicist and engineer.   A synthesis of the theory of differential systems, numerical analysis, and recursive functions is required: It will be called the Theory of Recursive Systems. Category theory is a useful mathematical setting: Beyond its evident foundational role in the theory of semantics of computer programs, it can serve as the underlying algebraic underpinning of computer science, much as linear algebra serves for classical Lie theory and differential geometry. Another mathematical goal is to generalize Lie theory itself to cover discrete situations, systems of difference equations and recursive functions After a review of some of the basic concepts of category theory, a discrete-time dynamical system is defined as a functor. The difference equations obtained as the one-step numerical analysis approximations of continuous-time dynamical systems are described in terms of cartesian products of two categories, one representing the 'dynamics', the other the 'step-size'. The 'step-size' parameter can be regarded as a 'deformation' parameter, much as the velocity of light in relativistic mechanics or Planck's constant in quantum mechanics. The one-step numerical approximations to continuous-time dynamical systems are extended from their traditional setting in Euclidean space to differentiable manifolds. The relations between Automata and Language theory (the latter considered algebraically as the theory of collections of subsets of free monoids) on the one hand, and feedback control theory on the other, are made more explicit. A beginning is made towards the goal of providing an appropriate geometrico-algebraic setting for a general theory of recursive and difference systems: The ultimate 'applied' goal is to provide a foundation for the theory of parallel and concurrent processes, systems and computer programs that is analogous to the foundations of the theory of von Neumann-type programs provided by the classical theory of recursive functions depending on one 'independent' variable. Thus the classical theory of recursive functions should stand in the same relation to a more general category-based theory of systems as the theory of 'integrable' ordinary differential equations stands to the general Ehresmann jet-space based theory of differential systems . The theory of feedback control systems also will take its natural place in this hierarchy. A theory of linear recursive systems is presented which parallels the theory of linear partial differential equations on manifolds.

Key words. Categories, sheaves, difference equtions, differential systems, deformation theory of algebraic and geometric structures, recursive functions, theory of computations, numerical analysis of ordinary differential equations, Lie group theory, control systems, automata, semigroups, monoid, computer science, programming theory

1. GENERAL INTRODUCTION

In the past thirty-five years, there has been a flowering of geometric mathematics, and its application in such disciplines as physics, mechanics, control theory, and logic. I suggest here that the natural progression will be towards the use of such geometric methods and ways of thinking as a unifying mathematical methodology in computing science.

This is the second work in a series [23] whose ultimate aim is the development of just such a 'geometric' computing science. I began studying the literature of computing science and logic five years ago with the aim of broadening the scope of control theory so that it can better interact with ideas under development in computing science. As I attempted to extend the point of view described in [23], I realized that a more comprehensive synthesis of contemporary geometric mathematics and computing science was required. The aim of this paper is to **begin** the job of providing this synthesis.

I will then give here a brief review of some of the more elementary algebraic principles underlying category theory and refer for more detail to the excellent introductory treatise [3] on "Category Theory and Computing Science" by Barr and Wells. Although I wrote this report before I was aware of this book, it turns out that the point of view towards category theory that I was aiming for has been brilliantly realized in their treatise. I also follow them in using the term 'computing science' as a term to describe the computation process as an entity abstracted from a particular machine or algorithm, while also using the traditional 'computer science' to refer to the broader combination of 'theory' and 'experiment'. This semantical point is important for my purposes because my goal is to understand the **mathematics** of computation rather than the 'engineering' details that are inevitably involved in writing efficient Languages and Algorithms. (Of course, I do not mean to denigrate the art of carrying out these details. I am a mathematician, not a hacker, although I admire the products of the hacker's art!)

As I mention in the Preface, the experience of the past thirty years seems to show that the initial ambitious AI program was (and probably remains!) out of reach: Evolution - and mathematics - goes from the simple to the complex. However, the ambitious 'Knwledge Representation' goals of AI research provide us with intellectual challenges. This paper suggests ideas for this task for one such area, the theory of differential systems, together with certain related areas of applied mathematics.

Category theory has been used successfully in computer science in the algebraic formulation of the logical foundations [16] and in the theory of semantics of computer programs [52]. Its use in this paper has a somewhat different motivation, more akin to the role it plays in algebraic geometry as an algebraic superstructure spanning both the 'discrete' and 'continuous' mathematical worlds.   So far, computer science has mainly been concerned with the 'discrete'. As it tries to move into the 'continuous' areas of science and engineering that are traditionally treated with differential equation methodology, it seems desirable to try to develop a mathematical formalism that will enable computing programs to operate in both worlds.

Most of the expositions of category and sheaf theory have been oriented towards pure mathematics. In the hope that they can be made more accessible and useful for 'applied' purposes, some of the beginning algebra of category theory is briefly reviewed here in a way that more closely resembles the sort of algebra (e.g., set theory, equivalence relations, partially ordered sets, matrix and group theory, etc.) already familiar to- and widely applied by- the community of mathematical engineers and physicists, rather than emphasizing the combinatorial and diagrammatic formulation that category theory inherited from its origins in algebraic topology.   The role that the category notion plays as a natural generalization of **both** the structures of 'partially-ordered sets' and 'groups' is particularly important.   Further, this unification serves the development of mathematical intuition - which is often 'geometric' in nature- since one sees the 'natural' and 'general' setting for many special constructions and proofs.   I will also review some links to the theory of Formal Languages, a topic that plays an important role in both the Wonham-Ramadge theory [57] and many computing science disciplines.

The theory of differential systems was created in the late 19th and early 20th centuries as a hybrid analytic and algebraic theory of systems of partial differential equations. The work of this classical period can be sampled in the treatises by Goursat [18], Janet [42], Riquier [59] and Thomas [68]. (A modern version of some of this material can be found in the treatise by Pommaret [56].) Unfortunately, much of this early work is inaccessible to contemporary mathematicians and computer scientists because of the changes in mathematical formalism and language which have taken place in the intervening years. This classical theory led - sometimes indirectly - to the following three disciplines, which **are** alive in the contemporary research world:

a) Elie Cartan's theory of exterior differential systems. [7, 25].

b) The theory of differential systems on jet bundles; the pioneering work due to C. Ehresmann, D. C. Spencer and H. Goldschmidt [9, 10, 17, 47, 66].

c) Differential algebra, begun in the 1920's by Ritt.

In historical parallel with this material - but with, as far as I am aware, no mutual communication - the theory of recursive functions [4, 60] was developed by logicians, and has turned out to be a main mathematical foundation of Computing Science. My goal is to develop a theory that will bring together the differential geometer's theory of systems of differential and difference equations and the logician's theory of recursive functions: I propose to call it the **theory of recursive systems.** I can only hope to make a beginning here at what I expect will be a long research road. There is also an associated 'deformation theory' that I can only begin to hint at here.

One possible starting point for this attempt at Grand Unification of Recursive Function and Differential System Theory is the observation that the classical theory of recursive functions resembles the theory of 'integrable' differential systems. In order to describe which problems are 'integrable' or 'solvable' or 'computationally feasible' it is essential to have a more comprehensive class of problems into which the specified class is embedded.

For differential systems, this more comprehensive framework is provided by the geometric theory of Ehresmann- Spencer jet spaces, bundles and sheaves. In this paper, I apply category and sheaf-theoretic ideas and methodology to the task of generalizing the theory of differential systems to cover the sort of recusively-defined 'discrete' systems encountered in computing science.

Sheaf theory was introduced in the 1950's in order to algebraically formalize topics encountered in topology and differential geometry. Of particular importance in its development was the geometric and algebraic reasoning underlying de Rham's Theorems about the relation between differential geometry and topology. D. C. Spencer and his co-workers pushed on from this beginning to explore the algebraic and geometric intricacies of the theory of systems of partial differential equations. One of my ultimate goals is to define a category and sheaf-theoretic based generalized differential system theory to apply to the systems occurring in computing science, and to extend the classsical Recursive Function Theory of Godel, Church and Turing in 'geometric' directions. I will try to keep in contact with applications, illustrating the general ideas with topics in control theory and the theory of systems of difference equations.

In [23], I pointed out some quasi-geometric and control-theoretic insights into the mathematical nature of a computer program. In order to motivate for control engineers some of the concepts introduced here, I will briefly describe some relations to Discrete Event System Theory [40, 57], a new branch of control theory which is better adapted to interaction with computing science. This material is also related to the Theory of Processes [21, 38, 53], a topic developed by computer scientists. Mathematically, the latter subject stands towards the former as an Axioms System of Logic stands towards its Models.

Over one hundred years ago, Sophus Lie began the theory of the interconnections between differential equations and group theory. He was in part motivated by what we would now call 'algorithms' for solving differential equations'. I believe that Lie's ideas - of course enriched with contemporary viewpoints - have a substantial role to play in development of algorithms for numerical analysis. Indeed, the numerical analysis of linear algebra   and ordinary differential equations [15, 22] should provide some of the most interesting applications for the formalism under development in this paper.   Certainly, it is the area of classical mathematics which   pre-eminently deals   with systems of difference equations!

Around thirty five years ago Charles Ehresmann constructed a mathematical formalism - the theory of jet spaces   - that opened the door to the codification of Lie's ideas about differential equations   in terms of twentieth-century mathematics.   Since then, there has been extensive development of the pure and applied aspects of the Lie theory of differential equations, but almost no work has appeared on the relation between this Lie theory and traditional numerical analysis.   The latter discipline developed from quite different roots than the former, with the pragmatic goal of constructing efficient algorithms of approximation   to solutions of differential equations.   These approximations are often constructed as solutions to difference equations.   However, there is not yet a Lie-Ehresmann theory of difference equations: In this paper I begin to show how one may be constructed, and used in the the study of approximations and in conventional numerical analysis.   I hope this will contribute to the development of numerical analysis along lines that make better use of the capabilities of both the modern theory of symbolic computation and Lie theory.   In previous work [29] I have begun developing such links between Lie theory and numerical   analysis.

I would like to thank Gerald Sussman and Dana Scott for guidance into the Brave New World of Computer Science.

2.   REMARKS ABOUT POSSIBLE LINKS BETWEEN   CATEGORY THEORY, CONTROL AND COMPUTER SCIENCE PROCESS THEORY

In the late 1950's,   mathematical system and control theory

[2, 44] crystallized out of a diverse background. The applied framework came mainly from the electrical engineering disciplines (control, circuit theory, estimation, cybernetics, automata,etc.), and the mathematics was that commonly used by engineers: Matrix theory, ordinary differential equations and the calculus of variations, probability and stochastic process theory, etc.. Mathematical System Theory served through the the '60s and well into the '70's as a 'theoretical' background for much 'practical' activity by a diverse group: Electrical, mechanical and chemical engineers, computer scientist

economists, etc. It also provided an interesting class of mathematical problems of considerable theoretical and applied importance, directly linked to some of the most 'advanced' and 'pure' mathematics of the day. The treatise by Caines [6] will give the reader a feeling for this diversity.

Although this System Theory has an intellectual structure which has much in common with that of Computer Science - particularly the Theory of Processes [21,38, 53] and Formal Language and Automata Theory [5, 8, 11, 19, 45, 46, 48, 54, 61] - unfortunately most of the problems of contemporary computing science can not be attacked with the same mathematical techniques. The theory of Discrete Event Systems of Ramadge and Wonham [57] is a pioneering effort to extend the ideas of System Theory to cover situations more closely linked to Computing Science.

As mentioned above, the differences between Process and Discrete Event System Theory reflect the difference in point of view between the computer scientist and the mathematical system theorist. Process Theory is an attempt to algebracize certain general constructions of computer science along the mathematical lines of Universal Algebra. The notion of a 'Process' is treated axiomatically, and the various ways Processes can be combined is described as a (free) algebra. Discrete Event System Theory uses the mathematical structures called 'automata' as a semantic model for a Process, in much the same way that the general algebraic phenomena dealt with in System Theory [44] may be modelled by difference and differential equations.

There is much common mathematical ground - particularly at the level of basic concepts and definitions - between 'control systems' as developed in the engineering and mathematics literature of the past fifty years and the theory of 'automata' as developed by logicians and computing scientists in the same period. While these links are more-or-less known to the specialists, the lack of an explicit exposition has held back communication between the diverse groups of people involved in research in this and neighboring areas. Although I cannot give a full-scale exposition in the limited space available in this paper, I will try to briefly indicate some of these relations, in the hope it will be helpful to the reader and. perhaps, stimulate a fuller treatement elsewhere.

Both System and Process Theory deal with mathematical structures that determine collections of 'data' . Computer Scientists call them 'traces', while in dynamical system and control theory they are called 'orbits' or 'trajectories'. In order to emphasize the links with dynamical system theory, I will use the term 'trajectories'.

In my previous paper [23], I have informally modelled such collections of data in a geometric way analogous to the methodology in the theory of Path Systems.   In this paper, I will model such structured data in the following general way: Let **C** be the category associated with a partially-ordered set, and let **SET** be the category of all Sets. A **Process** may be considered as a functor **F** from **C** to **SET**. A **trajectory** τ of the Process **F** will be defined as a functor from **C** to the category **PAIR** (whose objects are sets and whose morphisms are ordered pairs of sets) such that the image under τ of an object of **C** is an element of the image under **F** of the object and such that the corresponding morphisms fit together.   In this way, we obtain collections of 'points' constructed by a determined algebraic procedure from the data, represented by the functor **F**.   This is the algebraic mechanism underlying   such   assignments as :

Computing Science Processes --> Traces

Dynamical Systems --> Orbits or Trajectories

Differential Systems --> Sheaf of Solutions

Grammars --> Languages as Strings of Symbols of an Alphabet

Recursive Schemes --> Recursive Functions

Combining Category Theory and the existing Discrete Event System and Computing Science Process Theory will introduce unifying algebraic and geometric methodology to a discipline that is still in a formative stage. Most of the mathematical development in these disciplines has emphasized the syntactic component. Assigning models and semantics to syntactically-defined

Processes and Systems should certainly be conceptually clearer when processed through the categorical machinery . The category viewpoint might also be important when an attempt is made to extend the applications of Discrete Event System and Process Theory to such applied computing science areas as Artifical Intelligence.

3. INTERACTIONS BETWEEN LIE THEORY, DEFORMATION THEORY OF ALGEBRAIC STRUCTURES, AND THE NUMERICAL ANALYSIS OF ORDINARY DIFFERENTIAL EQUATIONS.

In the classical theory of numerical analysis of ordinary differential equations [15, 22] one attempts to solve, approximately, ordinary differential equations of the form:

$$dx/dt = f(x) \qquad (3.1)$$

$$x \, \varepsilon \, R^m$$

by means of a difference equation of the form:

$$x_n - x_{n-1} = h_{n-1} g(x_{n-1}, h_{n-1}) \qquad (3.2)$$

where:

$$t_0 < t_1 < \ldots < t_n < \ldots$$

$$t_n = t_{n-1} + h_{n-1}$$

is a discretization of the interval from t_0 to t. The positive real numbers h_n are the **step size**. In this paper, we we will deal only with situations of uniform step-size, which we will denote as follows:

$$h_n = h \text{ for all } n \qquad (3.3)$$

The standard example of such a .discretization is the **Euler method**:

$$x_n - x_{n-1} = f(x_{n-1})h. \tag{3.4}$$

The difference equation 3.4 is obtained from the integral equation

$$x_n - x_{n-1} = \int_{t_{n-1}}^{t_n} f(x(t))dt \tag{3.5}$$

- which is of course just the integrated form of the given differential equation 3.1 - by the simplest type of approximation of the integral on the right hand side of 3.5 and a uniform step-size 3.3.   More elaborate approximations of this integral - for example, using the classical poynomial interpolation formulas - lead to other difference equation approximations.

The difference equation 3.3 can be solved by iteration of an h-dependent map of R^n into itself.   Let us examine the structure of this solution.

Theorem 3.1.   For fixed step-size h, let:

$$\alpha(h): R^m \longrightarrow R^m \tag{3.6}$$

be the map defined by:

$$\alpha(h)(x) = x + f(x)h \tag{3.7}$$

Then:

The solution $n \longrightarrow x_n$ of the Euler difference equation 3.3 is given as the iteration of $\alpha(h)$, i.e.

$$x_n = \alpha(h)^n(x_0) \tag{3.8}$$

Suppose further that the map f: R^m --> R^m defining the differential equation 3.1 is continuously differentiable. Then, the solution

$$t \text{ --> } x(t)$$

of the differential equation 3.1 such that:

$$x(0) = x_0 \tag{3.9}$$

is approximated by x_n given by 3.8, with the following relation constraining values of 'step-size', 'length of iteration' and 'time':

$$hn = t \tag{3.10}$$

Lie group theoretically, this approximation of solutions of differential-equations by solutions of difference equations can be described in the following terms:

Let t --> g(t): R^n --> R^n be a one-parameter semigroup of diffeomorphisms of R^n generated by the ordinary differential equation 3.1, i.e. the orbit curves

$$t \text{ --> } g(t)x_0 = x(t) \tag{3.11}$$

are solutions of 3.1. Then,

$$g(t) = \lim_{n \text{ --> } \infty} \alpha(t/n)^n \tag{3.12}$$

where the limit on the right hand side of 3.12 is taken in the point-wise sense.

Proof. We can rewrite the difference equation 3.3 as follows:

$$x_{n+1} = x_n + f(x_n)h$$
$$= \alpha(h)(x_n) \tag{3.13}$$

Suppose that $(n, h) \longrightarrow x_n(h)$ is defined by 3.8. Then,

$$x_{n+1} = \alpha(h)^{n+1}(x_0) = \alpha(h)(\alpha(h)^n(x_0)) = \alpha(h)x_n, \tag{3.14}$$

which shows that 3.8 provides a formula for the solution of 3.3.

If $t \longrightarrow x(t)$ is the solution of 3.1, then the following formula follows from the standard results [15] on the Euler approximation:

$$x(t) = \lim_{n \longrightarrow \infty} x_n(t/n) \tag{3.15}$$

3.12 now follows from 3.11 and 3.15.

$$\mathbf{Q.E.D.}$$

We can generalize this Lie-theoretic approach to what Gear calls [15] the **one- step numerical analysis** of an ordinary differential equation. Here is one of his main qualitative results:

Theorem 3.2. Consider again the differential equation 3.1. Let

$$(x, h) \longrightarrow F(x, h) \tag{3.16}$$

$$R^m \times R \longrightarrow R^m$$

be a smooth map satisfying the following condition:

$$F(x, 0) = f(x) \tag{3.17}$$

Let $t \longrightarrow x(t)$ be a smooth curve in R^m that is a solution of 3.1 and let:

$$y: N \times R \longrightarrow R^m \tag{3.18}$$
$$(n, h) \longrightarrow y(n, h)$$

be the solution to the difference equation:

$$y(n+1, h) - y(n, h) = hF(y(n, h), h) \tag{3.19}$$

with the initial condition:

$$y(0, h) = x(0) \tag{3.20}$$

Then, for $t \in R$,

$$\lim_{n \to \infty} y(n, t/n) = x(t). \tag{3.21}$$

We can now indicate a Lie-theoretic interpretation of the key limiting relation 3.21.

Theorem 3.3. Let:

$$t \rightarrow g(t): R^m \rightarrow R^m$$

be the one-parameter pseudogroup of diffeomorphisms of R^m generated by the differential equation 3.1 , i.e. the orbit curves $t \rightarrow g(t)(x(0)$ of the pseudogroup are the solutions of 3.1 . Let F be a map with domain and range indicated in 3.16, satisfying relation 3.17. For $h \in R$, let:

$$\alpha(h): R^m \rightarrow R^m \tag{3.22}$$

be the map defined by the following formula:

$$\alpha(h)(x) = x + hF(x, h) \tag{3.23}$$

For each value of h, we define a monoid acting on R^m by the following formula:

$$n \rightarrow \alpha(h)^n \tag{3.24}$$

Then the family of transformation monoids defined by 3.24 approximates the one-parameter pseudogroup $t \rightarrow g(t)$ in the sense that the following relation is satisfied:

$$\lim_{n \to \infty} \alpha(t/n)^n(x) = g(t)(x) \text{ for all } x \in R^m. \tag{3.25}$$

Proof. Follows from 3.21, after identifying the left and right hand sides of 3.25 with the orbits of the semigroups appearing in 3.24 .

$$\textbf{Q.E.D.}$$

We can now extend this result to differential equations which live on an arbitrary differentiable manifold.

Theorem 3.4. Let X be a differentiable manifold, and let

$$t \longrightarrow \phi(t): X \longrightarrow X \qquad\qquad (3.26)$$

be a one-parameter Lie pseudogroup of diffeomorphisms acting on X. Let V be the vector field on X that is the infintesimal generator of the pseudogroup 3.26 , i.e. the following condition is satisfied:

> For each $x \in X$, the tangent vector $V(x) \in X_x$ $\qquad$ (3.27)
> which is the value of V at x is equal to the
> tangent vector to the orbit curve $t \longrightarrow \phi(t)(x)$
> at $t = 0$.

Let h denote a small real number. For each such h, let:

$$\alpha(h): X \longrightarrow X \qquad\qquad (3.28)$$

be a map. Suppose that the map: $(h, x) \longrightarrow \alpha(h)(x)$ is smooth. Suppose that the following conditions are satisfied:

> $\alpha(0) =$ the identity map $\qquad\qquad (3.29)$

> For each $x \in X$, the tangent vector to the curve $\qquad (3.30)$
> $h \longrightarrow \alpha(h)(x)$ at $h = 0$ is equal to $V(x)$

Then:

The one-parameter Lie pseudogroup 3.26 is approximated by the family:

$$\{ [\alpha(h)]^n: n = 0, 1, \dots \} \qquad\qquad (3.31)$$

of one-(discrete) parameter semigroups, in the sense that the following condition is satisfied:

For small t, $x \in X$,

$$\lim_{n \to \infty} \alpha(t/n)^n(x) = \phi(t)(x) \qquad (3.32)$$

Proof. By introducing a local coordinate system, we can reduce the general manifold case to R^m. Condition 3.30 - plus the smoothness hypotheses - guarantees that condition 3.23 is satisfied.

Q. E. D.

Remarks. Finding difference-equation approximations to differential equations is analogous to the problem of 'quantizing' classical mechanical systems. Intuitively, the relation between the difference equation 3.3 and the differential equation 3.1 may be thought of as replacing 'dx' by 'Δx', with:

$$[\Delta x](t, h) = [x(t+h) - x(t)]h^{-1} \qquad (3.33)$$

In geometric terms, 'dx' is an R^n - valued 1-differential form on R^n.

This material may also be usefully considered from the point of view of the Theory of Deformations of Algebraic Structures. The process whereby the difference equation 3.19 deforms to the differential equation 3.1
 as the step-size h goes to zero is analogous to the process in mathematical physics whereby Relativistic Mechanics goes over to Newtonian Mechanics, or Quantum Mechanics goes over to Classical Mechanics. The formalization and algebrazation of some of these ideas was developed in [31]. There are also probably important links to the theory of Lie semigroups, an important topic which finally now seems under development in the pure-mathematics literature [37].

Let us now make a brief excursion in the geometry of linear algebra numerical analysis.

4. A RIEMANNIAN - GEOMETRIC POINT OF VIEW IN LINEAR ALGEBRA NUMERICAL ANALYSIS.

There are many possible relations between Lie theory and that part of Numerical Analysis devoted to Linear Algebra.   As a by-product of my work [24, 25, 32, 33, 34, 35] in the 1970's on the Lie-theoretic and algebro-geometric foundations of linear system theory, (partially done in collaboration with Clyde Martin and also partially based on my other work [30] in the 1960's on harmonic analysis on symmetric spaces) I realized that similar methodology would be useful in numerical analysis in elucidating these relations. Since Martin expressed an interest in working with his student Greg Ammar on this material I passed the program on to them [1], after publishing the preliminary work that I had done in one of my books [29].

Recall the Lie-theoretic setting for the LR algorithm of numerical linear algebra. (It is typical of many other algorithms) Let G be the Lie group GL(n, C) of invertible, nxn matrices with complex coefficients. Let S be a maximal solvable subgroup of G, e.g. the subgroup of nxn matrices with zero elements above the diagonal. In Lie group theory, S is an example of what is called a Borel Subgroup, a type about which much is known. Given g ε G, a basic problem of numerical linear algebra is to find an element h ε G such that

$h^{-1}gh \ \varepsilon \ S.$

A natural geometric way to do this is to construct the coset space X = G/S, let g act on X via the left action of G, and look for a fixed point of this action. Now, from general Lie theory we know that X is compact and has positive Euler characteristic. Hence, the existence of such a fixed point follows from the Lefschetz Fixed Point Theorem. The problem for the Numerical Analyst is then to find a feasible algorithm for computing such a fixed point, exactly or approximately, subject to the usual constraints on the applied mathematician of Robustness, Efficiency, etc.

The simplest and most geometrically natural method for finding a fixed point is to find the closure of the subgroup generated by g. More concretely, start with a point $x \in X$, define the set

$$\{x_n: n = 0, 1, \ldots \}$$

inductively via the following formulas:

$$x_{n+1} = gx_n, \tag{4.1}$$

$$x_0 = x, \tag{4.2}$$

and try to find the fixed point as the limit as n goes to infinty of x_n.

Here is a general differential-geometric setting which is suggested by my previous work [30, 34] and, I believe, will serves as an intermediary between the 'concreteness' of the algorithms in the numerical analysis literature and the 'generality' of the algebraic-group literature.

Suppose that X is a submanifold of M, $x \in X$, $g \in G$, and:

$$gX \subseteq X \tag{4.3}$$

$$g = g'g'', \ g' \in G, \ g'' \in G \tag{4.4}$$

$$g'g'' = g''g' \tag{4.5}$$

Suppose further that :

There is a Riemannian metric on X $\qquad (4.6)$
such that g' is an isometry of that metric.

There is a real-valued function
f on X such that:

$$g'' = \exp(\operatorname{grad} f) \tag{4.7}$$

where f --> grad(f) is the gradient operator
on functions defined by the metric.

With these assumptions, one can discuss the local and global limiting behavior of the orbits of the cyclic group generated by g using general differential geometric and differential- topological technique. A beginning towards the study of such behavior can be found in my papers cited above.

5. THE 'FEYNMAN PRINCIPLE': A GENERAL LIE - THEORETIC PRINCIPLE IN THE THEORY OF APPROXIMATION OF SOLUTIONS OF DIFFERENTIAL EQUATIONS BY SOLUTIONS OF DIFFERENCE EQUATIONS

Let us continue the study of general Lie-theoretic principles hiding in the computational bushes of the discipline of Numerical Analysis.
Let X be a differentiable manifold; T   an interval of real numbers containing the point t = 0, and let

$$t \longrightarrow g(t), \quad t \; \varepsilon \; T \tag{5.1}$$

be a one - parameter pseudogroup of diffeomorphisms of X.   Let V be the vector field on X that is the infinitesimal generator of this one-parameter pseudogroup.   An **orbit** of the pseudogroup is a curve

$$t \longrightarrow x(t) , \; t \; \varepsilon \; T \; \text{in} \; X$$

that satisfies the following relation:

$$x(t) = g(t)(x(0)), \; t \; \varepsilon \; T \tag{5.2}$$

Such a curve satisfies the following ordinary differential equation:

$$dx/dt = V(x(t)) \tag{5.3}$$

The formula 5.2 describes what in the classical theory of ordinary differential equations would be called the **general solution** of the differential equation 5.3.

Many numerical analysis algorithms are based on splitting up the orbit curves of such a vector field into 'small' pieces, and then approximating each of these pieces by more computationally tractable curves. This is also a common idea in mathematical physics; most famously in the **Feynman Path Integral**. Hence, I will call it the 'Feynman Principle'. Here is a beginning towards a formalization of this process.

Theorem 5.1. Let $t \to g(t)$ be a one-parameter pseudogroup. Then, the following formula holds:

$$g(t) = g(a_n t + b_n) \cdots g(a_1 t + b_1) \tag{5.4}$$

where:

$$a_1, \cdots, a_n, b_1, \cdots, b_n$$

are real numbers satisfying the following relations:

$$a_1 + \cdots + a_n = 1 \tag{5.5}$$

$$b_1 + \cdots + b_n = 0 \tag{5.6}$$

Proof. 5.4 follows from iterating the defining property for the pseudogroup, namely:

$$g(a+b) = g(a)g(b)$$
if a, b and $a+b$ belong to the interval of
definition of the pseudogroup.

Q.E.D.

Theorem 5.2. The solution curve:

$$t \longrightarrow x(t) \tag{5.7}$$

to the ordinary differential equation 5.3 can be expressed in the following form:

$$x(t) = g(a_n t + b_n) \cdots g(a_1 t + b_1)(x(0)) \tag{5.8}$$

5.8 gives us a decomposition of the solution curve 5.7 into pieces of the following form:

$$x_1(t) = g(a_1 t + b_1)(x(0)) \tag{5.9}$$

$$x_2(t) = g(a_2 t + b_2)(x_1(t)) \tag{5.10}$$

$$\cdots$$

$$x_n(t) = g(a_n t + b_n)(x_{n-1}(t)) = x(t) \tag{5.11}$$

Proof. Use 5.4.

For t ranging over a fixed interval of real numbers, we can arrange - by appropriate choice of the a's and b's - that each piece 5.9-5.11 of the solution 5.7 involves the action of a 'small' element of the pseudogroup. In this way we can construct 'approximations' of the given solution of the ordinary differential equation using approximations given in advance for the orbits resuting from 'small' values of the group parameters.

Let us formulate this approximation 'philosophy' in the following qualitative way:

Theorem 5.3. Suppose that h: (t, x) ---> $h(t, x)$ is a local map of

$$T \times X ---> X \qquad (5.12)$$

satisfying the following condition:

For $x \in X$, t small, $h(t, x)$ is 'aproximately' $g(t)(x)$. $\qquad (5.13)$

Set:

$$y_1(t) = h(a_1 t + b_1, x(0)) \qquad (5.14)$$
$$y_2(t) = h(a_2 t + b_2, y_1(t)) \qquad (5.15)$$
$$\cdots$$
$$y(t) = h(a_n t + b_n, y_{n-1}(t)) \qquad (5.16)$$

Then, the curve

$$t ---> y(t) \qquad (5.17)$$

in X is an 'approximation' to the curve 5.2.

Proof That 5.17 is an 'approximation' of 5.7 follws from these formulas if the criterion of 'approximation' used has the following property:

If x is approximately x_1, and if t is small, then $g(t)(x)$ is approximately $h(t, x_1)$.

As an illustration of these general ideas, let us examine the Euler approximation mentioned above, starting from these group-theoretic first principles. First, use the group-property of the one-parameter pseudogroup $t ---> g(t)$:

$$g(t) = g(t/n)^n$$
$$= g(t/n) \cdots g(t/n) \qquad (5.18)$$

Hence,

$$x(t) = g(t)(x(0)) = x_n(t) = g(t/n)(x_{n-1}(t))$$
$$= g(t/n)g(t/n)x_{n-2}(t) = \cdots \qquad (5.19)$$

Let us use the following approximation, for n large:

$$g(t/n)(x) \cong h(t/n, x) \qquad (5.20)$$

Let suppose that the maps $(t, x) \longrightarrow g(t, x)$, $(t, x) \longrightarrow h(t, x)$, are smooth. One way to assure 5.20 is to assume that, for fixed x, the maps $t \longrightarrow g(t)(x)$ and $t \longrightarrow h(t, x)$ agree to a certain order at $t = 0$. The Euler method is associated with the following assumption:

> The maps $t \longrightarrow h(t, x)$ and $g(t)(x)$ agree to
> first order at $t = 0$. $\qquad (5.21)$

For the specific Euler method used in the numerical analysis literature, where the underlying manifold X is Euclidean space, acted on transitively by the affine Lie group, the approximating map

$$(t, x) \longrightarrow h(t, x)$$

has the following additional property:

> For each $x \in X$, there is a one-parameter pseudogroup
> of diffeomorphisms $t \longrightarrow k^x(t)$ such that $\qquad (5.22)$
> $h(t, x) = k^x(t)(x)$

The **Runge-Kutta** method similiarly involves curves $t \longrightarrow h(t, x)$ which agree with $t \longrightarrow g(t)(x)$ to higher order at $t = 0$.

6. A GEOMETRIC VIEWPOINT ON RICHARDSON EXTRAPOLATION

Continuing with our brief survey of Geometric principles in Numerical Analysis, let us turn to **Richardson Extrapolation,** another idea of classical Numerical Analysis. Here is a   brief explanation of its geometric essence.

Let X, H and Y be sets. Let:

$$f: X \times H \longrightarrow Y \qquad\qquad (6.1)$$
$$(x, h) \longrightarrow f(x, h)$$

$$g: X \longrightarrow Y \qquad\qquad (6.2)$$

be maps. In typical numerical analysis examples, the variable 'h' will be the 'step-size' of the 'discrete approximation f' of the problem whose 'exact' solution is 'g'. Thus, as h approaches some limiting value, $f(x, h)$ is to have something to do with g.

Let us suppose that one knows $f(x, h)$ at a sequence $h_0, \ldots, h_m$ of points. of   H such that the maps

$$x \longrightarrow f(x, h_i), i = 0, \ldots \qquad\qquad (6.3)$$

represent successively 'better' approximations to g.   Let

$$\rho: X \times H \cdots \times H \longrightarrow Y \qquad\qquad (6.4)$$

be a map. ρ wil be called a **Richardson Function** for the underlying numerical analysis problem if the composite function

$$x \longrightarrow \rho(f(x, h_0, \cdots, h_m) \qquad\qquad (6.6)$$

is a 'better' approximation to g than each of the maps h_i individually.

Remarks. In much of this material we deal again with what are geometrically-speaking deformations of differential equations by difference equations, thus coming within the domain of the general theory of **deformation of structures**. What seems to be involved is a theory of 'approximation' of one algebraic structure by another.

The Examples from Numerical Analysis just presented suggest the need for a general theory. Experience with other parts of mathematics suggests that Category Theory is an appropriate setting for such a theory. Let us then review some ideas from that branch of algebra.

7. THE DEFINITION OF A CATEGORY. TOPOLOGICAL AND DIFFERENTIAL STRUCTURES ON CATEGORIES

We have seen several examples of the approximation of one sort of algebraic structure by another, a general theme running through the discipline of numerical analysis. Let us begin the process of formalization of this idea in general terms. Let us start with a review of general definitions of category theory [3, 10, 12, 16, 20, 43, 49, 50, 51, 52, 55, 58], providing a setting for category theory in the form customarily chosen for the other algebraic structures of central importance in pure and applied mathematics, such as groups, rings and vector spaces. As we have mentioned above, the treatise by Barr and Wells [3] can be consulted for further detail.

Numerical Analysis also involves 'approximation' of differential equations by difference equations. In order to prepare for the formalization of what is meant by 'approximation' we describe one method of defining topological and differentiable spaces assigned to the algebraic structures called 'categories', in such a way as to be compatible with the standard notion of 'topological group' and 'Lie group'. These ideas are needed to make precise what is meant by 'approximation' for the categories that appear very naturally in Numerical Analysis.

One complicating feature in understanding the concept of 'category' is that it is a composite algebraic structure invoving two underlying sets and several operations interrelating them. We refer to the treatises by Barr and Wells [3] and MacLane [50] for the full story, recalling just enough here to fix notations.

Let us suppose given the following data:

a) Sets:

$$X = \{x, y, z, ..\} \text{ and } \Phi = \{\Phi\} \, ,$$

called the **objects** and **morphisms** (or **arrows**), respectively.

b) A **map from morphisms to pair of objects** :

$$\pi: \Phi \dashrightarrow X \times X. \tag{7.1}$$

For $(x, y) \, \varepsilon \, \pi(\Phi) \subset X \times X$, set:

$$\Phi(x, y) = \pi^{-1}((x, y)) = \{\Phi \, \varepsilon \, \Phi: \pi(\Phi) = (x, y)\} \tag{7.2}$$

For $\Phi \, \varepsilon \, \Phi$, the first component 'x' of $\pi(\Phi)$ is called the **source** of the morphism Φ, while the second component 'y' is called the **target** of Φ.

c) A **partially-defined binary product on morphisms**, satisfying the following conditions:

For each $(x, y, z) \, \varepsilon \, X \times X \times X$ such that, if

$$(x, y) \, \varepsilon \, \text{domain } \Phi \quad \text{and} \quad (y, z) \, \varepsilon \, \text{domain } \Phi,$$

then

$$(x, z) \, \varepsilon \, \text{domain } \Phi,$$

and there is given a map:

$$\Phi(x, y) \times \Phi(y, z) \dashrightarrow \Phi(x, z). \tag{7.3}$$

For notational simplicity, the map 7.3. is usually denoted as follows:

$$(\Phi', \Phi'') \dashrightarrow \Phi'\Phi'' \tag{7.4}$$

Let us suppose that these algebraic operations satisfy the following algebraic laws:

d). **Associative law for the morphism product**

For $(x, y, z, w) \in X \times X \times X \times X$,

$\Phi' \in \Phi(x, y)$, $\Phi'' \in \Phi(y, z)$, $\Phi''' \in \Phi(z, w)$, we have:

$$(\Phi'\Phi'')\Phi''' = \Phi'(\Phi''\Phi''') \qquad (7.5)$$

Finally, suppose that that this product has the following:

e) **Identity structure**

A map:

$$e : X \dashrightarrow \Phi$$

$$x \dashrightarrow e(x) \in \Phi(x, x) \qquad (7.6)$$

is given satisfying the following rules:

$$e(x)\Phi = \Phi \quad \text{for } \Phi \in \Phi(x, y) \qquad (7.7)$$

$$\Phi e(x) = \Phi \quad \text{for } \Phi \in \Phi(y, x). \qquad (7.8)$$

Definition. The composite algebraic structure:

$$C = (X, \Phi, \pi, \alpha, e) \qquad (7.9)$$

satisfying relations a) - e) is called a **category**.

For notational simplicity, we shall often denote the category by the labels given to its objects and morphisms, as :

$$C = (X, \Phi) \text{ or } C(X, \Phi) \qquad (7.10)$$

Remark. As has already been noted, the algebraic structure described by the category concept plays a basic role in Computer Science. In addition to the place of category theory in Semantics and Logic, another reason for this is that - as we shall review in the next Sections - the category axioms 7.1-7.8 reduce to those of a pre-ordered set and a monoid when a special choice is made for the morphisms and objects. Thus, in a sense the first-order algebraic structure called a 'category' [20] interpolates between the two first-order algebraic structures - 'partally-ordered sets' and 'monoids' - which have already proved to be basic for Computer Science research. For example, partally-ordered sets are the algebraic structures involved in the Scott-Strachey theory of Denotational Semantics [52, 62, 63, 64, 65, 68, 70], while monoids are the primary algebraic structures occurring in Formal Language Theory [5, 8, 11, 13, 14, 19, 39, 45, 46, 48, 54, 61]. As Ehresmann realized in the 1950's, category theory is also a fundamental algebraic structure in the modern description of Lie theory.

Example. The category N of non-negative integers.

The object set of N is the set N of non-negative integers. The morphism set of N is the collection of ordered pairs

$$(m, n) \; \varepsilon \; N \times N \text{ such that } m \leq n.$$

The product on these morphisms is defined as follows:

$$(m, n)(p, q) \text{ is defined if and only if:}$$

$$p = n, \tag{7.11}$$

and then:

$$(m, n)(n, q) = (m, q) \tag{7.12}$$

Remark. We often use the notational convention inherent in this example: The category is denoted by the same label as its objects, but in bold-face type.

Example.   The category of real matrices.

The object set is the set N. For each

$$(m, n) \; \epsilon \; N \times N, \tag{7.13}$$

let the set of morphisms whose source object is "m" and target object is "n" be the set of m x n real matrices. The partially-defined product on morphisms is matrix multiplication. The identity element assigned to each object "n" is the n x n identity matrix.

Example. The category whose morphism set is identical to its object set, called the   Category of Constants

Let X be a given set. We will now define a category - denoted as

$$C(X, X) \tag{7.14}$$

whose objects and morphisms are both identical to X. Let:

$$\pi: X \dashrightarrow X \times X \tag{7.15}$$

be the diagonal map:

$$\text{morphisms} \dashrightarrow \text{objects} \times \text{objects} \tag{7.16}$$

defined by the following formula:

$$\pi(x) = (x, x) \text{ for } x \; \epsilon \; X. \tag{7.17}$$

π identifies the object and morphism sets. The morphism product is defined as follows:

For x, y ε X, xy is defined if and only if:

$$x = y, \tag{7.18}$$

If (7.18) is satisfied, then:

$$xx = x. \tag{7.19}$$

The Identity structure is defined as follows:

$$e(x) = x \tag{7.20}$$

All the conditions needed to define a category are satisfied.

Remark. This category plays a basic - but unheralded - role in both physics and numerical analysis situations. In the former discipline, it is one of the building blocks of 'gauge fields', while for the latter it is -as we shall see - the algebraic structure involved in the 'step size' formalism used for discretization of differential equations.

Example. The direct (or cartesian) product of two categories.

Let $C(X, \Phi)$, $C(X', \Phi')$ be two categories. The **direct (or cartesian) product** of them is the category $C''(X'', \Phi'')$ whose objects and morphisms are defined as follows:

$$X'' = X \times X' \tag{7.21}$$

$$\Phi'' = \Phi \times \Phi' \tag{7.22}$$

$$\pi''(\phi \times \phi') = (\pi(\phi) \times \pi'(\phi')) \tag{7.23}$$

The partially-defined product operation on Φ'' is defined as follows:

$$(\phi_1, \phi'_1)(\phi_2, \phi'_2) \text{ is defined if and only if:}$$

$$\phi_1 \phi_2 \text{ and } \phi'_1 \phi'_2 \text{ are defined,} \qquad (7.24)$$

and

$$(\phi_1, \phi'_1)(\phi_2, \phi'_2) = (\phi_1 \phi_2, \phi'_1 \phi'_2) \qquad (7.25)$$

Again, the reader will readily verify that that $C''(X'', \Phi'')$ - with these operations - is a category.

Let us now turn to the question of imposing toplogical structures on categories, a topic of importance in understanding ideas of Numerical Analysis.

Definition. Let

$$C = (X, \Phi, \pi, \alpha, e) \qquad (7.26)$$

be a category, as defined above. A **topological structure** for C is defined by giving the following data:

> Topological structures in the usual sense for X and Φ,
> i.e. collections of subsets for X and Φ (to be called the
> **open sets**) which are closed under the operations of
> finite intersection and abitrary union $\qquad (7.27)$

subject to the following conditions:

$$\pi \text{ is a continuous map: } \Phi \rightarrow X \times X, \qquad (7.28)$$

where the topology on X x X is the standard product topology.

> The partially defined morphism multiplication maps

$$\Phi \times \Phi \rightarrow \Phi \qquad (7.29)$$

are continuous, in the standard product and subset topologies.

The map $e: X \to \Phi$ is continuous. $\qquad\qquad$ (7.30)

The reader can check that natural topologies - induced from the standard topology on the real numbers - may be imposed on the categories described in the above examples.

We can now extend these definitions to cover the notions of 'differentiability', 'smoothness' and 'topological structure' defined on categories. Following the pattern used to define the notion of a 'Lie group' one requires that the spaces involved be 'differentiable manifolds' and that the algebraic operations be 'differentiable' with respect to this manifold strucrure. Many of the categories we shall encounter can be identified with subsets of real Euclidean space, hence will inherit topologies and manifold structures from the standard ones on real Euclidean space.

8. SETS WITH PRE-ORDERS AS THE OBJECT SETS FOR CATEGORIES

Since partially-ordered sets are of such great importance in computer science, let us make explicit their place in category theory.

Definition. Let X be a set. A **relation** on X is a subset R of $X \times X$. A **pre-order** on X is a relation $R \subset X \times X$ satisfying the following conditions:

a) If $(x, y) \in R$ and $(y, z) \in R$, then $(x, z) \in R$. $\qquad$ (8.1)

b) $(x, x) \in R$ for all $x \in X$. $\qquad\qquad$ (8.2)

Let us use the following more familiar notation:

$(x, y) \in R$ if and only if $x \leq y$. $\qquad\qquad$ (8.3)

Axioms a) and b) translate into the following:

a') $x \leq y$ and $y \leq z$ imply $x \leq z$ **(Transitivity)**   (8.4)

b') $x \leq x$ for all $x \in X$.   (8.5)

If the pre-order relation $R = \{\leq\}$ satisfies the following additional condition

c) $x \leq y$ and $y \leq x$ implies $x = y$ (8.6)

then $\{\leq\}$ is called a **partial ordering** and X is called a **partially-ordered set** or **poset.**

Let $\{X, \leq\}$ denote a fixed pre-order relation on the set X.. We shall construct a category $C(X, \leq)$ from it:

The 'objects' of $C(X, \leq)$ are the elements of X.   (8.7)

The 'morphisms' of $C(X, \leq)$ are the set of ordered pairs

$(x, y) \in X \times X$

such that:

$x \leq y$ (8.8)

The map π from the space of morphisms to the cartesian product of the object space with itself is defined as:

$\pi((x, y)) = (x, y)$ (8.9)

The partially-defined product mapping of

(morphisms) x (morphisms) --> (morphisms)

is defined as:

If $x \leq y$ and $z \leq w$, then $(x, y)(z, w)$ is defined
if and only if:

$$y = z, \qquad (8.10)$$

and then:

$$(x, y)(z, w) = (x, w)) \qquad (8.11)$$

The identity-element structure $x \dashrightarrow e(x)$ is defined as :

$$e(x) = (x, x) \text{ for } x \, \varepsilon X. \qquad (8.12)$$

Theorem 8.1. With the structure $C(X, \leq)$ assigned as above to a pre-order $\{X, \leq\}$, $C(X, \leq)$ satisfies the category axioms. Thus, a 'pre-order on a set X' is a category whose morphism set is identified with a subset of $X \times X$, and whose object space is X itself.

Proof. A routine verification, left to the reader. Note that Axiom a) is essentially the associative law for morphism multiplication.

9. MONOIDS AND MONOIDS-WITH-GAUGE -STRUCTURE AS CATEGORIES.

A 'monoid' is an algebraic structure which is defined analogously to a 'group', without the 'existence - of - inverses - axiom', but with the 'associative law' and a single 'inverse element'. In other words, a 'semigroup' with an 'identity element'. Here is the formal

Definition. Let M be a set. A **monoid** structure on M is defined by a mapping:

$$M \times M \longrightarrow M,$$

$$(m_1, m_2) \longrightarrow m_1 m_2 \tag{9.1}$$

satisfying the following axioms:

Associative Law: $\quad m_1(m_2 m_3) = (m_1 m_2)m_3 \tag{9.2}$

for $m_1, m_2, m_3 \ \varepsilon \ M$.

Identity: There is an element $e \ \varepsilon \ M$ such that

$$em = me = m \tag{9.3}$$

for all $m \ \varepsilon \ M$.

(One can then prove that 'e' is unique).

Example. The monoid of real, square matrices.

Let n be an integer and let M be the set of n x n matrices with real coefficients. Let the binary product: M x M --> M be matrix multiplication. The identity element is the identity matrix. With these definitions, the 'monoid' conditions 9.2-9.3 follow from properties of matrix multiplication.

Here is a general setting:

Theorem 9.1. Let M be a monoid. It defines a category, which we shall denote by **M**. There is but one object, the Identity Element 'e' of the monoid structure. The morphism set is **M** itself. The multiplication :

$$morphisms \ x \ morphisms \ \text{-->} \ morphisms \qquad (9.4)$$

is that given by the monoid axioms.

Proof. It should be obvious that **M** so defined satisfies the category axioms.

Remark. A **group** is then a monoid such that every element has an inverse. A **semigroup** is what is left of the 'monoid' structure after the 'existence-of-identity' axiom is thrown away, i.e. a set together with a binary algebraic structure satsifying the associative law.

Example. Monoids with gauge structure as categories.

In general, as for any algebraic structure, 'complicated' categories can be built up from 'simple' ones, by operations, e.g. cartesian product. Here is an illustration of this principle.

Let M be a monoid, denoted multiplivatively with identity element e, and let H be a set. Consider the following partially-defined product on H x M:

For h, h' ε H, m, m' ε M, the product
of (h, m) and (h', m') is defined if and only if:

$$h = h',$$
(9.5)

and then:

$$(h, m)(h, m') = (h, mm').$$
(9.6)

Theorem 9.2 The partially-defined product on H x M defined by 9.6 defines a category denoted as **H x M** whose object set is H x {e}. The morphism set assigned to a pair (h, e) and (h',e) of objects is the empty set, unless 9.6 is satisfied. The morphisms set assigned to the pair {(h, e), (h,e)} of objects is the set H. The identity element corresponding to the object (h, e) is (h, e) itself.

Proof. The structure defining **H x M** is clearly the cartesian product of the category $C(H, H)$ defined by 7.14-7.19 and the category **M** associated with the given module structure on M. In particular, it satisfies the category axioms 7.1-7.9.

10. THE CATEGORY 'SET', 'PAIR' AND 'RELATION'.

As for any general mathematical structure, category theory was abstracted from many diverse examples and specialized situations that play an important role in various branches of mathematics. One of these 'examples' is the theory of sets and mappings. We adopt the 'naive, working-mathematician's' point of view towards set-theory, ignoring the famous Paradoxes of the Foundations, such as the difficulties surrounding the use of such terms as 'the set of all sets'. We only deal with restricted, more-or-less constructible pieces of this vast domain, although it would unduly complicate the exposition to be precise at each stage.

Definition. Let **SET** be the categorical structure with the following 'objects' and 'morphisms':

The objects of **SET** are sets. Denote a typical object by 'A'.

The morphisms of **SET** are maps between sets. Explicitly, if

$$\phi: A \dashrightarrow B \tag{10.1}$$

is a map with domain A and range contained in B, then the ordered triple:

$$\{\phi, A, B\} \tag{10.2}$$

is an element of the morphism space Φ.

The map:

$$\pi: \text{morphisms} \dashrightarrow (\text{objects}) \times (\text{objects})$$

is defined as follows:

$$\pi(\{\phi, A, B\}) = (A, B) \tag{10.3}$$

The partially-defined map:

$$\text{morphisms} \times \text{morphisms} \dashrightarrow \text{morphisms}$$

is defined as follows:

$$\{\phi, A, B\}\{\phi', A', B'\} \text{ is defined if and only if:}$$

$$B' = A, \tag{10.4}$$

and then:

$$\{\phi, A, B\}\{\phi', A', B'\} = \{\phi\phi', A', B\}, \qquad (10.5)$$

where " $\phi\phi'$ " is the usual "composition of maps", read as:

$$\text{First apply } \phi', \text{ then apply } \phi . \qquad (10.6)$$

The Identity Rules for **SET** are as follows:

> Associate with the set A, i.e. the object of **SET,** (10.7)
> the identity map of A to itself.

The reader will readily verify that **SET** defined in this way is a category.

Remark. In category theory it is customary to reverse the notation for composition of maps, in order to have other formulas come out looking more elegant. However, as one who has spent twenty years teaching calculus, I dislike changing the classical meaning of $z(y(x))$:

$$\text{First find } y(x), \text{ then apply the function 'z' to it,} \qquad (10.8)$$

or to attempt to use some variant notation, such as:

$$(x)y$$

Each branch of mathematics seems to deal with categories which are subcategories of **SET** attached to certain basic algebraic structures. For example, Number Theory and Classical Recursive Function Theory deal with categories of mappings between sequences of integers. Much of numerical analysis deals with categories of polynomial maps between sequences of real or complex numbers. Classical algebraic geometry deals with a category whose objects are 'algebraic' subsets of complex Euclidean or projective spaces and whose morphisms are 'algebraic' maps between them. (It is difficult to be precise about the meaning of the term 'algebraic' without going into the technicalities of commutative algebra.) A recent paper by Blum, Shub and Smale [5] highlights the role that categories consisting of polyomial maps between 'semi-algebraic' sets play in the theory of computation and complexity. Here is a version of the category with which they deal.

Example. The category of polynomial maps of finite dimensional free modules over commutative, associative rings.

Let CR be a commutative associative ring. For each positive integer n, CR^n denotes the set of n-tuples of elements of CR. A map:

$$\phi: CR^n \dashrightarrow CR^m \tag{10.9}$$

is a **polynomial map** if the components of f are 'polynomial functions' in the traditional sense in terms of the ring CR. We are again confronted with a category which is a subcategory of **SET**. The objects are the integers, and the morphism set attached to (m, n) the collection of poynomial maps $: CR^n \dashrightarrow CR^m$.

Here is another category that appears often in mathematical and computer science practice:

Example. The category PAIR (of sets).

The category **SET** defined above has a combinatorial structure that is independent of the details of the maps. One aspect of this structure may be captured by constructing another category that we call **PAIR**.

The objects of **PAIR** are sets, typically denoted as 'A'. $\tag{10.10}$

The morphisms of **PAIR** are ordered pairs (A, B) of objects. $\tag{10.11}$

The partially-defined product on morphisms is defined as follows:

$$(A, B)\, (A', B') \text{ is defined if and only if}$$

$$B' = A$$

and   then

$$(A, B)\, (A', B') = (B', A) \tag{10.12}$$

Let us now extend the category **SET** by adding elements which represent relations between sets.

Definition. Let A and B be sets. A **relation** with **domain A and codomain B** is a subset **R** of the Cartesian product A x B.

Let us now define an operation of **composition** between two relations whose domains and codomains match up. Suppose that:

$$\mathbf{R} \subset A \times B \tag{10.13}$$

$$\mathbf{R}' \subset B \times C \tag{10.14}$$

are two such relations. We will define a product $\mathbf{R}'\mathbf{R}$ as follows:

$$\mathbf{R}'\mathbf{R} \text{ is a relation with domain A and codomain C} \tag{10.15}$$

$$\mathbf{R}'\mathbf{R} = \{(a, c): a \,\varepsilon\, A, b \,\varepsilon\, C, \text{ and there exists a } b \,\varepsilon\, B \tag{10.16}$$
$$\text{such that } (a, b) \,\varepsilon\, \mathbf{R} \text{ and } (b, c) \,\varepsilon\, \mathbf{R}'\}$$

Theorem 10.1. The partially-defined binary operation

$$(R', R) \dashrightarrow R'R \tag{10.17}$$

defined above defines a category **REL** whose objects are the objects of **SET** and whose morphisms are relations between sets.

Proof. One must prove that the partially-defined binary operation $(R', R) \dashrightarrow R'R$ satisfies the associative law. This is a straightforward compution which is left to the reader.

Remark. The point of view in the book by Eilenberg and Elgot [12] is also important and useful. They remark that a relation $R \subset A \times B$ defines a map:

$$\text{subsets of } A \dashrightarrow \text{subsets of } B, \tag{10.18}$$

hence the category **R** may also be considered as a subcategory of the category **SET**.

11. FUNCTORS BETWEEN CATEGORIES AS THE ANALOGUE OF 'HOMOMORPHISMS'. DISCRETE DYNAMICAL SYSTEMS AS FUNCTORS

In the set-theoretic codification of algebra, (for example, the treatises of van der Waerden and Lang) a 'homomorphism' of an 'algebraic structure' is a 'structure-preserving' map between two sets that carry an example of the 'algebraic structure'. Since the founders of category theory were abstracting features that appear naturally in the topological situation, they gave the name **functor** to what in the van der Waerden - style would be called 'homomorphism'.

Definition. Let $C = (X, \Phi)$ and $C' = (X', \Phi')$ be two categories. A **functor** F from C to C' is defined by a pair of maps (χ, δ):

$$\chi: X \dashrightarrow X' \tag{11.1}$$

$$\delta: \Phi \dashrightarrow \Phi' \tag{11.2}$$

satisfying the following conditions:

$$\delta(\Phi(x, y)) \subset \Phi'(\chi(x), \chi(y)) \text{ for } x, y \in X. \tag{11.3}$$

$$\delta(\Phi_1 \Phi_2) = \delta(\Phi_1)\delta(\Phi_2) \tag{11.4}$$

for $\Phi_1, \Phi_2 \in \Phi$ such that the product is defined.

$$\delta(e(x)) = e'(\chi(x)) \text{ for } x \in X. \tag{11.5}$$

C is called the **source**, C' the **target** category. We will often abbreviate the notation and denote the functor as follows:

$$F(C) = C' \tag{11.6}$$

or

$$F : C \dashrightarrow C' \tag{11.7}$$

Example. The functor from SET to PAIR

Let **SET** and **PAIR** be the categories defined in Section 10.

Theorem 11.1. Define two mappings as follows:

$$\chi : \text{objects of } \textbf{SET} \dashrightarrow \text{objects of } \textbf{PAIR} \tag{11.8}$$

$$\delta : \text{morphisms of } \textbf{SET} \dashrightarrow \text{morphisms of } \textbf{PAIR} \tag{11.9}$$

satisfying the following conditions:

$$\chi \text{ is the identity map on the objects of } \textbf{PAIR} \tag{11.10}$$

δ maps a morphism of **SET**, i.e. a map

$$\phi : A \dashrightarrow B \tag{11.11}$$

into the Pair (A, B).

Then, the pair (χ, δ) of maps satisfies conditions 11.3 required to define a functor from the category **SET** to the category **PAIR**.

Proof. Routine verification, left to the reader.

A type of mapping between categories occurs often which satisfy the above axioms, with the exception that the image of a product is the product of the images in **reverse** order. Here is the relevant:

Definition. Let $C = (X, \Phi)$ and $C' = (X', \Phi')$ be two categories. A **contravariant functor** F^* from C to C' is defined by a pair of maps (χ, δ):

$$\chi: X \dashrightarrow X' \tag{11.12}$$

$$\delta: \Phi \dashrightarrow \Phi' \tag{11.13}$$

satisfying the following conditions:

$$\delta(\Phi(x, y)) \subset \Phi'(\chi(y), \chi(x)) \text{ for } x, y \; \varepsilon \; X. \tag{11.14}$$

$$\delta(\Phi_1 \Phi_2) = \delta(\Phi_2)\delta(\Phi_1) \tag{11.15}$$

for $\Phi_1, \Phi_2 \; \varepsilon \; \Phi$ such that the product is defined.

$$\delta(e(x)) = e'(\chi(x)) \text{ for } x \; \varepsilon \; X. \tag{11.16}$$

Again, C is called the **source**, C' the **target** category of the contravariant functor.

Remark. If F^* is a contravariant functor from C to C' it is also a functor from the 'dual' category C^* to C'. (C^* **is** obtained from C by 'reversing the direction of the arrows'.)

Let us now see how these ideas can be used to algebracize standard concepts from numerical analysis.

Example: The first order difference equations that are the generators of discrete one=parameter flows as functors from N to SET.

THEOREM 11.2. The collection of functors between the categories **N** and **SET** is in one-one correspondence with the sequences $\{X_n, f_n: n \in N\}$, where:

$$\text{Each } X_n \text{ is an object of } \mathbf{SET} \qquad (11.17)$$

$$\text{Each } f_n \text{ is a map: } X_n \dashrightarrow X_{n+1}, \text{ i.e., a morphism of } \mathbf{SET}. \qquad (11.18)$$

We can associate with each sequence $\{f_n: n \in N\}$ of maps of form 11.18 the following difference equation:

$$x_{n+1} = f_n(x_n) \qquad (11.19)$$

to be solved for sequences:

$$\{x_n: n \in N, x_n \in X_n\} \qquad (11.20)$$

Proof. Using the above definitions, we see that a functor from the category **N** to the category **SET** is determined by giving the following data:

$$\text{A map } n \dashrightarrow X_n \text{ from N to the objects of } \mathbf{SET}. \qquad (11.21)$$

For each pair $(m, n) \in N \times N$ such that $m \leq n$, a map

$$f_{m, n}: X_m \dashrightarrow X_n \qquad (11.22)$$

such that, whenever

$$m \leq n \text{ and } n \leq p, \qquad (11.23)$$

we have:

$$f_{m,\,p} = f_{n,\,p}f_{m,\,n} \tag{11.24}$$

Using 11.24 iteratively, we see that $f_{m,\,n}$ is the result of composition of maps of the form $f_{m,\,m+1}$.

Q.E.D.

Example.   Ordinary difference equations that depend on parameters as functors from the Cartesian product of N and the category of constants C(H, H).

In Section 3, we have seen that the one-step numerical approximation of ordinary differential equations leads to difference equations of the following form:

$$x_{n+1} = f_n(x_n,\, h) \tag{11.25}$$

where h is an element of a set H. Let us associate this sort of difference equation - which of course also occurs in Control Theory - with a functor between two categories.

Theorem 11.3.   Let:

$$\textbf{f: N x C(H, H) --> SET} \tag{11.26}$$

be a functor betwen the categories indicated in 11.26. Then, **f** is determined by giving the following data:

$$\text{A set } X(n,\, h) \text{ for each } (n,\, h)\ \varepsilon\ N \text{ x } H \tag{11.27}$$

$$\begin{gathered}\text{A map: } f_{n,h}\colon X(n,\, h) \text{ --> } X(n+1,\, h)\\ \text{for each } (n,\, h)\ \varepsilon\ N \text{ x } H\end{gathered} \tag{11.28}$$

In particular, the difference equations of the form 11.25 are special types of functors of the form 11.26.

Proof.   The morphisms of the category $N \times C(H, H)$ are the pairs of the form:

$$\{(n, h), (m, h): n \leq m; n, m \in N; h \in H\} \tag{11.29}$$

An arbitrary morphism of the form 11.29 is the composition of morphsims of the form:

$$\{(n, h), (n+1, h): n, \in N, h \in H\} \tag{11.30}$$

Hence, a functor from $N \times C(H, H)$ to **SET** is determined exactly as described in   11.27-11.211.

Q.E.D.

Example. Subcategories

Whenever mathematicians encounter 'algebraic structures', they automatically think of 'substructures'.   (Examples: 'subgroups' 'subspaces', 'subrings', etc.)   Applied to 'categories', this leads to the 'subcategory' concept, which is also an example of a 'functor':

Definition.   Let $C = (X, \Phi)$ be a   category. A pair (X', Φ') consisting of subsets:

$$X' \subset X \tag{11.31}$$

$$\Phi' \subset \Phi \tag{11.32}$$

is said to be a **subcategory** of C if the following condition is satisfied:

$$\text{a). } \Phi'\Phi' \subset \Phi', \tag{11.33}$$

i.e. the product of two morphisms of Φ' is again contained in Φ' whenever it is defined in terms of the category C

b). The identity elements of **C** attached to the objects   (11.34)
   in X' are identity elements of **C'**.

With the structure inherited from **C**, **C'** then becomes a category in its own right, and the inclusion map **C** --> **C'** is a functor.

Example. If X is a set, the collection of maps from X to X forms a Subcategory of **SET**, that we call **MAP(X)**. It is, in fact a monoid, since it contains only one object, namely the Identity Map of X into itself.

12. TRAJECTORIES. SPECIALIZATION TO DIFFERENTIAL AND DIFFERENCE EQUATIONS.

We can now use these concepts to formalize the 'geometric' notion of the 'set of solutions' (or 'trajectories') of.difference equations. Examples have been given in Section 8. Let us first introduce a general concept.

Definition. Let C and **C'** be categories, with **C'** a subcategory of the category **SET**. Let **F** be a functor from C to **C'**. A **trajectory,** denoted by τ, of the functor **F** is a functor from C to the category **PAIR** that satisfies the following conditions:

For each object x of C, $\tau(x)\ \varepsilon\ \mathbf{F}(x)$ (12.1)

For each object x of C, the image object $\tau(x)$ belongs to $\mathbf{F}(x)$.   (12.2)

For each pair (x, y) of objects of C and each morphism ϕ of C
   with source x and target y, the image set-map $\mathbf{F}(\phi)$ between
   sets satisfies the following condition:

$$\mathbf{F}(\phi)(\tau(x)) = \tau(y)$$ (12.3)

Theorem 12.1. A **solution** or **trajectory** in the classical sense of the difference equation 11.19 is a functor $\tau : \mathbf{N} \dashrightarrow \mathbf{PAIR}$ that is a trajectory of the functor $\mathbf{F} : \mathbf{N} \dashrightarrow \mathbf{SET}$ associated with the difference equation 11.21, as described in Theorem 11.2.

Proof. Let

$$\tau : \mathbf{N} \dashrightarrow \mathbf{PAIR} \qquad\qquad (12.4)$$

be a functor that is a Trajectory of the functor $\mathbf{F}$. It follows from the Definitions above that the following statement holds:

> The image object under $\mathbf{F}$ of the object n of $\mathbf{N}$
> is the set X_n that is involved in the definition $\qquad (12.5)$
> of the difference equation 11.19.

Let x_n be the image object of $\mathbf{PAIR}$ of the object n under τ . Condition 12.2 is equivalent to the following:

$$x_n \; \varepsilon \; X_n. \qquad\qquad (12.6)$$

The image under $\mathbf{F}$ of the morphism $(n, n+1)$ of $\mathbf{N}$ is the map:

$$f_n: X_n \dashrightarrow X_{n+1} \qquad\qquad (12.7)$$

Condition 12.3 implies:

$$f_n(x_n) \; = \; x_{n+1}, \qquad\qquad (12.8)$$

i.e., $n \dashrightarrow x_n$ is a solution of the difference equation 11.19. The proof of the converse, that each solution of 11.19 defines a Trajectory of $\mathbf{F}$, follows from the reversibility of these arguments.

Q.E.D.

Example. Difference equations with parameters

As we have seen, in order to represent and numerically approximate ordinary differential equations it is necessary to consider difference equations with parameters of the following form.

$$x(n+1) = f_n(x(n), h) \tag{12.9}$$

In Section 8, we have seen one way to represent such a gadget in terms of functors. Let us now see how the solution of 12.9 can be described as Trajectories. Let us again make the following assumptions:

$$n \text{ is an object of the category } \mathbf{N}, \tag{12.10}$$

$$x(n) \; \varepsilon \; X_n, \text{ an object of the category } \mathbf{Set}. \tag{12.11}$$

$$h \; \varepsilon \; H, \text{ an object of the category } \mathbf{Set}. \tag{12.12}$$

$$f_n \text{ is a map: } X_n \times H \dashrightarrow X_{n+1}. \tag{12.13}$$

Let us associate in the following way a functor $\mathbf{F} : \mathbf{N} \dashrightarrow \mathbf{SET}$ with this data:

$$\mathbf{F} \text{ maps the object n of } \mathbf{N} \text{ into the set } X_n \times H, \text{ an object of } \mathbf{SET} \tag{12.14}$$

$$\mathbf{F} \text{ maps the morphism } (n, n+1) \text{ of } \mathbf{N} \text{ into the following map:}$$

$$\mathbf{F} \; (n, \; n+1)(x_n, h) = (f_n(x_n), h) \tag{12.15}$$
$$X_n \times H \dashrightarrow X_{n+1} \times H$$

We can define $\mathbf{F} \; (n, n+k)$ for $k > 0$ by recursion on k:

$$\mathbf{F} \; (n, n+k) = \mathbf{F} \; (n+k-1, n) \; \mathbf{F} \; (n, n+k-1). \tag{12.16}$$

Theorem 12.2 Let **F** be the functor from **N** to **SET** defined by 12.13-12.15. Then, the trajectories of this functor are the maps:

$$n \dashrightarrow (x_n, h_n) \, \varepsilon \, (X_n, H) \qquad (12.17)$$

satisfying the following conditions:

$$x_{n+1} = f_n(x_n, h_n) \qquad (12.18)$$

$$h_{n+1} = h_n. \qquad (12.19)$$

Such a trajectory can then be identified with a solution of the difference equation 12.9.

Proof. Follows from 12.16.

Difference equations in which one 'independent variable' changes at a time provides the 'dynamics' corresponding to the 'von Neumann' type of Computer Program. The following example begins the extension of the ideas developed above in the direction of 'parallel' programs.

Example. A pair of difference equations in two independent variables. The difference equation analogue of Frobenius integrability as the condition for functorality.

We have seen that the simplest sort of first-order difference equation in one independent integer variable leads to interesting category-theoretic material. Let us turn to the next-simplest classical example, a system of two first-order difference equations in two independent variables. Consider one of the following form:

$$x(j+1, k) = f_{j, k}(x(j, k)) \tag{12.20}$$

$$x(j, k+1) = g_{j, k}(x(j, k)) \tag{12.21}$$

Consider N x N as a partially-ordered set which is the direct product of the standard partial ordering on N:

$$(j, k) \leq (j', k') \text{ iff } j \leq j' \text{ and } k \leq k' \tag{12.22}$$

N x N, with this partial ordering, is the object set of a category which is denoted as **N x N**.

Consider as given the following data:

$$x(j, k) \, \varepsilon \, X(j, k), \text{ an object of } \mathbf{SET} \tag{12.23}$$

$$(j, k) \, \varepsilon \, N \times N. \tag{12.24}$$

(j, k) --> $X(j, k)$ is a map:

$$\text{objects of } \mathbf{N \times N} \text{ --> objects of } \mathbf{SET} \tag{12.25}$$

$$f_{j, k} \text{ is a map: } X(j, k) \text{ --> } X(j+1, k) \tag{12.26}$$

$$g_{j, k} \text{ is a map: } X(j, k) \text{ --> } X(j, k+1) \tag{12.27}$$

Let us find the conditions that the following map:

$$x(j, k) \text{ --> } X(j, k) \tag{12.28}$$

on objects can be extended to be a functor from **N x N** to **SET**.

Theorem 12.3. There is a functor $F : N \times N \longrightarrow \mathbf{SET}$ such that:

$$F(j, k) = X(j, k) \qquad (12.29)$$

$$F((j, k), (j+1, k)) = f_{j, k} \qquad (12.30)$$

$$F((j, k), (j, k+1)) = g_{j, k} \qquad (12.31)$$

if and only if the following conditions are satisfied:

$$g_{j+1, k} f_{j, k} = f_{j, k+1} g_{j, k} \qquad (12.32)$$

Proof. Relations 12.30 and 12.31 determine the value of the functor F on the morphisms

$$((j, k), (j+1, k)) \qquad (12.33)$$
$$(j, k), (j, k+1) \qquad (12.34)$$

of the category $N \times N$. A morphism of $N \times N$ is a pair of pairs of non-negative integers:

$$\{(j, k), (j', k')\} \qquad (12.35)$$

satisfying the following conditions:

$$j \leq j' \text{ and } k \leq k' \qquad (12.36)$$

Hence:

$$\{(j, k), (j+1, k+1)\} \text{ is a morphism of } N \times N \qquad (12.37)$$

Also,

$$\{(j, k), (j+1, k+1)\} = \{(j, k), (j+1, k)\}\{(j+1, k), (j+1, k+1)\} \qquad (12.38)$$

where the product on the right hand side of 12.38 is the morphism product in the category $N \times N$. We can also write the left hand side of 12.38 in the alternate order as follows:

$$\{(j, k), (j+1, k+1)\} = \{(j, k), (j, k+1)\}\{(j, k+1), (j+1, k+1)\} \qquad (12.39)$$

Equating the right hand sides of 12.38 and 12.39 gives the following equality:

$$\{(j, k), (j+1, k)\}\{(j+1, k), (j+1, k+1)\}= \{(j, k), (j, k+1)\}\{(j, k+1), (j+1, k+1)\}$$
$$(12.40)$$

Apply F to 12.30 and 12.31 to obtain relation 12.32. This finishes the proof of the fact that 12.32 is a necessary condition for the existence of the functor F.

Let us prove the converse. We can write a morphism $\{(j, k), (j+m, k+n)\}$, $m, n > 0$, of $N \times N$ as a product of 'elementary' morphisms which only involve one step:

$$\{(j, k), (j+m, k+n)\} = \{(j, k), (j+m-1, k+n)\}\{(j+m-1, k+n), (j+m, k+n)\}$$
$$(12.41)$$

and so on. For each morphism $\{(j, k), (j+m, k+n)\}$ we can thus define the image morphism $F\{(j, k), (j+m, k+n)\}$ as a product of maps of the form of the right hand side of 12.30 and 12.31 by choosing a sequence of points in the morphism space going from each $\{(j, k), (j, k)\}$ to $\{(j, k), (j+m, k+n)\}$ involving elementary steps of the from 12.33-34. It only remains to show that the resulting morphisms $F\{(j, k), (j+m, k+n)\}$ satisfy the associative law and the identity law when they are multiplied together. This is readily done using the conditions 12.40 recursively. **Q.E.D.**

Remarks. The fact that the functor F exists if conditions 12.40 are satisfied is basically a cohomology argument that can be described using standard homological algebra machinery . Conditions 12.40 may also be interpreted as **integrability conditions** [17, 64] for the system of difference equations 12.20-12.21 associated with the functor **F**.

13. CATEGORIES WHOSE OBJECTS ARE SYSTEMS OF ORDINARY DIFFERENTIAL EQUATIONS AND WHOSE MORPHISMS ARE PROLONGATION MAPS. FEEDBACK CONTROL SYSTEMS AS EXAMPLES.

As we have seen in [23], some computer programs can be mathematically modelled by discrete-time feedback control systems. Continuous time feedback control systems are a special type of differential system. This Section reviews certain features of differential system theory from the point of view of the connection to the theory of feedback control systems. We deal here with the theory of differential systems, with one independent variable, which we usually denote by 't', and physically think of as 'time'. We will also assume some familiarity with the notations of the Ehresmann theory of jets of mappings between differentiable manifolds.

Consider a system of ordinary differential equations (say p of them), of the following form:

$$f(t, \, dx/dt, \, \ldots, \, dx^r/dt^r) = 0 \qquad\qquad (13.1)$$

where $x \in R^m$ and $t \in R$. The differential equation 13.1 is determined by the smooth map:

$$f: R^{m(r+1) + 1} \dashrightarrow R^p \qquad\qquad (13.2)$$

appearing on the left hand side of 13.1. A **solution** of 13.1 is a smooth map

$$t \dashrightarrow x(t) \qquad\qquad (13.3)$$

of an interval of R to R^m which makes the left hand side be equal to zero when it is substituted into 13.1.

We identify the domain space $R^{m(r+1) + 1}$ on the left hand side of 13.1 with the space $J^r(R, R^m)$ of r-jets of maps of R to R^m and interpret f as a map:

$$f: J^r(R, R^n) \rightarrow R^p \qquad (13.4)$$

Given a smooth map x: $t \rightarrow x(t)$ of an interval of R to R^n, the point

$$(t, dx/dt, ..., dx^r/dt^r) \; \varepsilon \; J^r(R, R^n) \qquad (13.5)$$

is denoted as:

$$j^r(x)(t) \qquad (13.6)$$

where x denotes the map and $j^r(x)$ denotes its r-jet. Set:

$$DE = f^{-1}(0) \qquad (13.7)$$

We see that a map x: $t \rightarrow x(t)$ is a solution of the differential equation 13.1 if and only if:

$$j^r(x)(t) \; \varepsilon \; DE \text{ for all } t \; \varepsilon \text{ domain } x. \qquad (13.8)$$

These ideas suggest a jet- theoretical definition of what is meant by a 'differential equation', a 'solution', and - most important from the categorical point of view we are trying to develop here - a 'morphism'.

Defintions. Let X be a (finite dimensional, paracompact, infinitely differentiable) manifold. An **r-th order differential equation system with configuration space** X **and independent variable space** R is a submanifold DE of $J^r(R, X)$. A **solution** of the differential equation is a smooth map:

$$x: R \rightarrow X \qquad (13.9)$$

such that:

$$j^r(x)(R) \subset DE \qquad (13.10)$$

Let us now construct a category whose objects are the differential systems and whose morphisms are 'prolongations' between differential systems, in the sense first used by Vessiot and Cartan in their classical work on differential systems and Lie pseudogroups. Suppose given two differential equation systems:

$$DE \subset J^r(R, X), DE' \subset J^{r'}(R, X') \tag{13.11}$$

with the same independent variable space R.

Definition. A **morphism** (or **prolongation**) from the unprimed to the primed system indicated in 13.11 is a smooth map:

$$\phi : DE \dashrightarrow DE' \tag{13.12}$$

which satisfies the following conditions:

If **x**: R --> X is a solution of 13.10, then there is a map **x'**: R --> X' such that

$$j^r(\mathbf{x'})(R) \subset DE' \tag{13.13}$$

and

$$j^r(\mathbf{x'})(t) = \phi(j^r(\mathbf{x})(t)) \text{ for } t \; e \; R. \tag{13.14}$$

Theorem 13.1. There is a category - a subcategory of **SET** - whose 'objects' are the differential systems and whose 'morphisms' are the differential system morphisms as defined above.

Proof. It must be verified that the composition of two differential system morphisms is again a differential system morphism. This is left to the reader.

Example. **The prolongation which converts the 'independent' varable'into a 'dependent variable'.**

This construction is a standard one in the 19th century literature. e.g. the 'phase portrait' construction of Poincare for a scalar ordinary differential equation of the form:

$$dy/dx = f(y, x); \quad x, y \; \varepsilon \; R. \tag{13.15}$$

Namely, set:

$$z = (y, x) = (z_1, z_2) \; \varepsilon \; R^2 \tag{13.16}$$

Then,

$$dz/dx = (dy/dx , 1) = (f(y, x), 1) = (f(z_1, z_2), 1) \tag{13.17}$$

is an ordinary differential equation for the vector valued function of x. The cartesian projection map:

$$(z_1, z_2) \; \text{-->} \; z_1 = y \tag{13.18}$$

then obviously maps solutions of 13.17 into solutions of 13.15, i.e. is a 'prolongation' as defined above. The generalization to ordinary differential equation systems of the form 13.1 should be obvious to the reader.

Let us turn now to another sort of 'prolongation', differentiating both sides of a differential equation.

Theorem 13.2.   Suppose given an r-th order differential system on R^n, given in the form 13.1, in terms of a map of the form 13.3.   Then, there is a map:

$$f^{(1)}: J^{r+1}(R, R^m) = R^{n(r+2)+1} \longrightarrow R^p \qquad (13.19)$$

such that:

$$d/dt[f(t, dx/dt, \ldots, dx^r/dt^r)] = f^{(1)}(t, dx/dt, \ldots, dx^{r+1}/dt^{r+1}) \qquad (13.20)$$

for each map $t \longrightarrow x(t)$.

Let $DE^{(1)}$ be the (r+1)-st order differential system defined by the maps f and $f^{(1)}$ from $J^{r+1}(R, R^m)$ to R^p.   Then, the canonical projection map

$$J^{r+1}(R, R^m) \longrightarrow J^r(R, R^m) \qquad (13.21)$$

(i.e. the map resulting from 'forgetting' the highest derivative) is a prolongation map of the system $DE^{(1)}$ to the system DE.

Proof.   The existence of $f^{(1)}$ satisfying 13.20 - as well as an explicit formula - follows from the 'chain rule' of differential calculus. The second part follows from the definitions.

$$\textbf{Q.E.D.}$$

In 19th century differential system theory - for example, in Sophus Lie's theory of contact transformations and in the theory of Baecklund transformations - one finds an extended notion of 'morphism', analogous to the extension of the 'single-valued function' concept to 'many-valued functions'.   In the set-theoretic formulation of the 'function' concept, this is of course obtained by replacing the standard definition of a function in terms of ordered pairs by the notion of **relation**.   Carried over to differential equations, Lie and Cartan suggested various notions of 'equivalence'. Let us briefly review one of their main ideas.

Definition. Let DE and DE' be systems of differential equations. An **extended morphism** between them is defined by a triple:

$$(DE'', \phi, \phi') \tag{13.22}$$

where:

$$DE'' \text{ is a third differential system.} \tag{13.23}$$

$$\phi: DE'' \dashrightarrow DE \tag{13.24}$$

$$\phi': DE'' \dashrightarrow DE' \tag{13.25}$$

are morphisms of differential systems. A pair $(\mathbf{x}, \mathbf{x}')$ of solutions of DE amd DE' are said to **correspond** under the extended morphism (DE'', ϕ, ϕ') if there is a solution $\mathbf{x}''$ of DE'' such that $\phi(\mathbf{x}'') = \mathbf{x}$ and $\phi'(\mathbf{x}'') = \mathbf{x}'$.

Remark. This construction has an abstact, category-theoretic foundation. Let $\mathbf{C}$ be a category. An **extended morphism** of $\mathbf{C}$ is a pair (ϕ, ϕ') of morphisms of $\mathbf{C}$ which have a common source object.

Another type of differential system morphism is traditionally called a 'subsystem'.

Definition. Let $DE \subset J^r(R, X)$ and $DE' \subset J^r(R, X)$ be differential systems. DE is said to be a **subsystem** of DE' if every solution of DE is also a solution of DE'.

We can now illustrate some of these concepts by indicating the structure of some of the systems of ordinary differential equations that are of importance in the the branch of engineering called feedback control theory:

Example.   Feedback control systems as differential systems

Let us now specialize to the following system of ordinary differential systems frequently encountered in engineering practice [41].

$$\frac{dx}{dt} = f\ (x,\ u) \tag{13.26}$$

$$y = g\ (x) \tag{13.27}$$

$$x \in R^m, u \in R^p, y \in R^q \quad .$$

13.26-13.27 is called a **feedback control system.**   u is called the **control,** or **input,** x the **state,** y the **output.**   Such a system is also called an **input-output system in state space form.**   The **open-loop** or **input-output** interpretation of this system is that an input curve $t \to u\ (t)$ is given,   the differential equations (13.26) are solved to obtain the state curve $t \to x\ (t)$, with initial condition

$$x\ (0) = x_o, \tag{13.28}$$

and the output $t \to y\ (t)$ is now obtained by substituting the state curve into the output map (13.21).   The result is to define a map

$$\text{(curves in U) x } X \to \text{(curves in Y)} \tag{13.29}$$

that is called the **input-output system** associated with (13.26) — (13.27), and denoted in block-diagram form as follows:

$$
\begin{array}{c}
u \longrightarrow \boxed{\ \ x\ \ } \longrightarrow y
\end{array}
$$

$$\tag{13.30}$$

Let us convert the feedback control system denoted by 13.26-13.27 into a differential system of the type we are considering in this Section. Set:

$$z = (x, u, y) \qquad (13.31)$$

Then,

$$dz/dt = (dx/dt, \ du/dt, \ dy/dt)$$

$$=, \text{ using } 13.25, \ (f(x, u), \ du/dt, \ dy/dt) \qquad (13.32)$$

Theorem 13.2. Let z denote the vector of R^{m+p+q} given in terms of the (state, input, output) by formula 13.31: Let (π, π', π'') be the following cartesian projection maps:

$$\pi(x, u, y) = x, \ \ \pi: R^{m+p+q} \to R^m \qquad (13.33)$$

$$\pi'(x, u, y) = u, \ \ \pi': R^{m+p+q} \to R^p \qquad (13.34)$$

$$\pi''(x, u, y) = y, \ \ \pi'': R^{m+p+q} \to R^q \qquad (13.35)$$

In terms of these maps, the state-space Control equations 13.26-13.27 take the following differential system form:

$$\pi(dz/dt) - f(\pi'(z)) = 0 \qquad (13.36)$$

$$\pi''(z) - g(\pi(z)) = 0. \qquad (13.37)$$

Proof. Follows on substituting 13.35-13.36 into 13.25-13.26.

Q.E.D.

Remark. In [27], I showed the value of this geometric, path-system point of view (classically, called **Monge Systems**) in studying the control systems in 'state space' form, namely 13.26-.27.. Putting the 'state' and 'control' vectors together in one big space also has geometric and Lie-theoretic value

in understanding the geometric foundations of feedback stabilization theory
[33, 35, 24]

The engineer's concept of a feedback control law may be interpreted
geometrically as a subsystem of the differential system (13.31) — (13.32)
associated with an open-loop feedback control system. This is the geometric
version of a way of looking at 'feedback' in the Wonham-Ramadge [55]
theory of control of Discrete Event Systems. The engineer's starting point for
thinking about feedback is the following diagram:

$$(13.38)$$

The intuitive idea is that the "signal" u enters the "black box" labelled as "x,"
emerges as "output" y, and then is "fed back" via the loop in (13.38) to be
incorporated into a "new" input u. This can be implemented in a standard
mathematical way as follows: Add to the differential equations
(13.25) — (13.27) determining the input-output conditions the relation

$$u = K(y) \qquad . \tag{13.39}$$

The result is to define a subsystem of the differential system
(13.26) — (13.27) determined by the following differential equations:

$$\frac{dx}{dt} = f(x, u)$$
$$y = g(x) \tag{13.40}$$
$$u = K(y) \qquad .$$

This system takes its most familiar and well-studied form for the
special case of a linear system. Let us sum up for this most important special
case in the following way:

Theorem 13.3. Consider a linear (possibly time-varying) system of the following form:

$$\frac{dx}{dt} = A(t)\, x + B(t)\, u$$

(13.41)

$$y = C(t)\, x$$

$$x \in R^n, u \in R^m, y \in R^p \quad .$$

The coefficients $t \to A(t)$, $B(t)$, $C(t)$ are curves in matrix spaces of the appropriate sizes. Consider linear output feedback of the following form:

$$u = K(t)\, y \quad ,$$

(13.42)

where $t \to K(t)$ is a curve in the space of $m \times p$ matrices. The equations resulting from "closing the loop" in the input-output system (13.41) are the following linear differential equations:

$$\frac{dx}{dt} = (A + B\,K\,C)\, x \quad .$$

(13.43)

Then, for each solution $t \to \hat{x}\,(t)$ of the differential equation (13.43), there is a solution

$$t \to \hat{u}\,(t), \hat{x}\,(t), \hat{y}\,(t)$$

(13.44)

of the open-loop system equations given by the following formulas:

$$\hat{y} = C(t)\, \hat{x} \quad ,$$

(13.45)

$$\hat{u} = K(t)\, C(t)\, \hat{x} \quad .$$

(13.46)

Proof. (13.43) follows from (13.41) by substituting (13.42) into (13.41). The second part of the theorem follows by observing that, given $\hat{x}$ as a solution of 13.43), relations (13.44) and (13.45) define a solution of the open-loop system equations (13.41).

Q.E.D.

Remark. One of the most important topics in the theory of linear feedback control is the relation between the existence of a stabilizing linear (state) feedback law and controllability of the system itself. This relation was first developed by R. Kalman in the 1950's: Indeed, his results in this area - and in the 'dual' Estimation and Filtering problems - constitute one of the highlights of the historical development of mathematical system theory. Kalman used a mixed analytic and algebraic argument to prove his results. (This argument is given in [2] in a form most accessible to a mathematician interested in the geometric side of Control Theory.) As a by-product of our research in the mid-1970's on the application of differential geometry and Lie group theory to control system theory, C. Martin and I gave another proof [34, 36] of Kalman's feedback stabilization results (and partially extended them to the 'output feedback situation) by means of tools of Lie theory and algebraic geometry. I believe that our methods - particularly if they can be made to work better in the 'discrete' and 'nonlinear' world of Computer Programs - will be useful in Computer Science, since - as I indicated in [23] - the theoretical aspects of the development of a computer programming strategy involve something very similiar to the engineer's 'closed-loop feedback stabilization problem'.

14. CATEGORIES WHOSE OBJECTS ARE SYSTEMS OF DIFFERENCE EQUATIONS. EXAMPLES FROM NUMERICAL ANALYSIS AND RECURSIVE FUNCTION TEORY

Let us return to the general theory of differential and difference systems. We will extend the standard Ehresmann-Spencer jet space ideas from 'differential' to 'difference' systems by following the classical numerical analysis idea of replacing 'derivatives' by 'differences'. We will also use the definition of the 'primitive recursive functions' [4, 60] as a guide to our development.

Let us begin with the difference equation analogue of the system 13.1 of ordinary difference equations, modified to include 'extra parameters', as in the theory of recursive functions [60]. Let:

$$n \to X(n) \tag{14.1}$$

be a map from the object set of the category **N** to the object set of the category **SET**. Let A be a set, which will serve as a parameter set. Consider a system of ordinary difference equations of the following form:

$$f(n, x(n+r, a), x(n+r-1, a), \ldots, x(n, a), a) = y_0 \tag{14.2}$$

where:

$$a \, \varepsilon \, A \tag{14.3}$$

$$x(n) \, \varepsilon \, X(n) \text{ for } n \, \varepsilon \, N. \tag{14.4}$$

For each n and r ε N, set:

$$X(n, r) = \{(n, x(n+r), \ldots, x(n), a): n \, \varepsilon \, N, \ (x(n+r), \ldots, x(n)) \, \varepsilon \, X(n+r) \times \ldots \times X(n), a \, \varepsilon \, A\} \tag{14.5}$$

f is a given map with the following domain and range:

$$f: X(n, r) \to Y \tag{14.6}$$
$$y_0 \, \varepsilon \, Y$$

Computer scientists often refers to 14.2 as a 'recursion' , perhaps with the map 'f' not given explicitly. We can reformulate the meaning of 14.2 to a form which is more convenient for computer science purposes. For each n, set:

$$DS(n) = \{z \in X(n, r): f(z) = y_0\} \qquad (14.7)$$

That the map $(n, a) \rightarrow x(n, a)$ is a solution of the difference equation 14.2 means that:

$$(n, x(n+r, a), x(n+r-1, a), ..., x(n, a), a) \in DS(n) \qquad (14.8)$$
$$\text{for all } (n, a)$$

Let us now use this reformulation to provide a general definition in set-theoretic terms:

Definition. Let r be a non-negative integer. With the above notations, a **difference system of order** r (with the category $\mathbf{N}$ as 'independent variables') associated with this data is defined by a family of sets:

$$n \rightarrow DS(n)$$

parameterized by $n \in N$ such that:

$$DS(n) \subset X(n, r) \qquad (14.9)$$

A map: $(n, a) \rightarrow x(n, a)) \in X(n)$ is a **solution** of the difference system:

$$\{X(n), n \in N, DS(n), A\} \qquad (14.10)$$

if the following condition is satisfied, for all $n \in N, a \in A$:

$$(x(n, a), ..., x(n+r, a)) \in DS(n) \qquad (14.11)$$

A **morphism** between the difference systems $\{X(n), n \in N, DS(n)\}$ and $\{X'(n), n \in N, DS'(n)\}$ is defined by a map:

$$\text{Object set of } \mathbf{N} \rightarrow \text{morphism set of } \mathbf{SET}$$

$$n \rightarrow \phi(n) \qquad (14.12)$$

such that the following conditions   are satisfied:

$$\phi(n): DS(n) \dashrightarrow DS'(n) \tag{14.13}$$

If n --> x(n) is a solution of {DS(n)}
then there is a solution t -->x'(n) of {DS'(n)}
such that:

$$(x'(n), ..., x'(n+r)) = \phi(n)((x'(n), ..., x'(n+r)) \tag{14.14}$$

for all $n \in N$.

Remark. The simplest case is that where all the {X(n)} are identified with a single set X. Let X^r be the Cartesian product of r copies of X. Then, X^{r+1} plays the same role that the jet space $J^r(R, X)$ does in the theory of differential systems.

Example. The difference equations appearing in the classical theory of primitive recursive functions.

Let X and A be sets. The theory of 'primitive' recursive function deals with maps

$$x: A \times N \dashrightarrow X \tag{14.15}$$

defined by recursion relations of the following type:

$$x(a, n+1) = F(a, x(a\ n)) \tag{14.16}$$

were F is a given (e.g. possibly by a previous application of the recursion rules) map of the following domain and range:

$$F: A \times X \dashrightarrow X. \tag{14.17}$$

Remark. Note the resemblance between the equations 14.16 and the difference equations of one-step aproximations to continuous-time dynamical systems reviewed in Section 3.   In this case, the parameter space A is identified with the 'step-size' space, denoted by 'H' in Section 3.

Let us now formulate a point of view towards systems of difference equations of the general type 14.2, where we regard each such system as a functor from the category **N** to the category **REL**.   For simplicity, consider a difference system of the following form:

$$F_n(x(n+1, a), x(n, a), a) = y_0 \qquad (14.18)$$

$$x(n, a) \; \varepsilon \; X(n) \qquad (14.19)$$

$$a \, \varepsilon \, A$$

Remark.   A general system of type 14.2 can be prolonged to one of type 14.18.

In 14.19, $n \dashrightarrow X(n)$ is a map:

$$\mathbf{N} \dashrightarrow (\text{objects of } \mathbf{SET})$$

and

$$F_n\colon X(n+1) \times X(n) \times A \dashrightarrow Y \qquad (14.20)$$

is a map with indicated domain and range.   We can associate with the difference equation 14.18 the subset:

$$DE(n) \subset X(n+1) \times A \times X(n) \times A \qquad (14.21)$$

defined by the following formula:

$$DE(n) = \{(x(n+1), a, x(n), a))\colon F_n(x(n+1), a, x(n), a)) = y_0\} \qquad (14.22)$$

Theorem 14.1. Consider each DE(n), for n ε N, as a (binary) relation with domain X(n) x A and codomain X(n+1) x A. Then, there is a unique functor:

$$N \longrightarrow REL \qquad\qquad (14.23)$$

such that the following conditions are satisfied:

The object n of N goes into the object X(n) x A of **REL**. $\qquad (14.24)$

The morphism (n, n+1) of **N** maps into the relation
DE(n) defined by 14.22, considered as a morphism $\qquad (14.25)$
of **REL**.

Thus, the set of difference equations of type 14.18 can be parameterized by the set of functors **N** --> **REL**.

Proof. For each integer k > 0, assign to the morphism

$$(n, n+k) \qquad\qquad (14.26)$$

of **N** the product within **REL** of the morphisms corresponding to (n+k-1, n+k), ... , (n, n+1). This argument proves that the functors **N** --> **REL** are determined uniquely by the maps of form 14.22.

Q.E.D.

15. TRANSFORMATION ACTIONS OF MONOIDS, AUTOMATA AND FEEDBACK CONTROL

In [23], it was pointed out that concepts of feedback control theory seemed to play a hidden role in computer program theory. One source for this connection between Computer Science and Control Theory is their common ancestry in the 1940's and 1950's: 'Automata' and 'feedback control systems' are variants of the same basic idea. However, it is difficult to find recent material explicitly describing this connection. I will briefly describe a few such relations here.

The algebraic structure that underlies both Automata and Control Theory is also that which motivated the work of Sophus Lie in the 19th century on what he called 'transormation groups'. Lie at first tried to make a general theory about the structure of what we would now call 'semigroups of diffeomorphisms of differentiable manifolds'. He ran into trouble since he could say very little of great significance about the 'infinitesimal' structure of such objects. However, when he added the assumption that the semigroup was actually a 'group', and that the transformations of the group were generated locally as solutions of differential equations, he hit pay-dirt. His pioneering ideas were picked up by others - especially Killing, E. Cartan and Weyl - and the resulting theory of Lie Groups is now a major part of pure and applied mathematics.

The modern version of Lie's original concept can be described as follows:

Definition. Let M be a monoid, and let X be a set. An **action of M by transformations on X** is defined by a map:

$$M \times X \dashrightarrow X \tag{15.1}$$

$$(m, x) \dashrightarrow mx,$$

such that the following conditions are satisfied:

$$m(m'x) = (mm')(x) \tag{15.2}$$

$$ex = x \tag{15.3}$$

for $m, m' \in M$, $x \in X$, where 'e' is the identity element of the monoid M.

Given such an action, for each $m \in M$, define a map

$$f(m): X \longrightarrow X$$

as follows:

$$f(m)(x) = mx \quad \text{for } x \in X. \tag{15.4}$$

The map $m \longrightarrow f(m)$ is a functor between the category associated with the monoid M and the category associated with the monoid **MAP(X)**.

One major difference between the material in the Automata and Control literature is that the former mainly deals with monoids which are free in the algebraic sense, while the latter deals with monoids which may have a non-free structure. For example, for the most successful Control theory - that associated with Linear Systems - the monoids are submonoids of the monoid of affine maps of a vector space into itself, which is very far from being free! In what is now called 'Geometric Control Theory', one deals with monoids generated by diffeomorphisms of differentiable manifolds, with special attention to those generated as orbits of smooth vector fields.

Let us recall a few concepts that play a major role in the Automata literature.

Definition. Let M be a monoid. M is said to be **free** if the following condition is satisfied:

For each finite subset $\{m_1, \ldots, m_n\}$ of M

$$m_1 \cdots m_n = e \tag{15.5}$$

if and only if:

$$m_1 = e, \ldots m_n = e$$

Definition. Let M' be a subset of a monoid M. M' is said to be a **submonoid** of M' if $e \in$ M', and $m_1 m_2 \in$ M' for m_1, $m_2 \in$ M'. The intersection of a collection of submonoids is again a submonoid. If V is an arbitrary subset of M, the **submonoid generated by V, denoted as** V^*, is the intersection of all submonoids of M containing V.

Definition. Let M be a monoid. A **language L associated with M** is subset L of M. If L is such a language, a **sublanguage** of L is a subset L' of L.

The aim of much of Formal language and Automata theory is to describe how languages are recursively generated from certain subsets of free monoids by means of 'grammers' and 'production rules'. The standard treatises [5, 19, 46, 56] on Formal Language Theory deal at considerable length with ways that languages can be generated by Automata, Turing Machines, etc. Here are some algebraic concepts that appear in this work.

Definition. Let M be a monoid and let $(M_1, \ldots, M_n)$ be an ordered n-tuple of subsets of M. Set:

$$M_1 \cdots M_n = \{m_1 \cdots m_n : m_1 \in M_1, \ldots, m_n \in M_n\} \tag{15.6}$$

$M_1 \cdots M_n$ is then the subset of 'words' obtained from the 'words' of $M_1, \ldots, M_n$ by stringing them all together.

Theorem 15.1. Let M be a monoid. Let $L(M)$ be the set of sublanguages of M. Define two binary algebraic operations on $L(M)$:

$$L \times L' \dashrightarrow L \cup L' \tag{15.7}$$

$$L \times L' \dashrightarrow L \cdot L' \tag{15.8}$$

These two operations define a type of algebraic structure on $L(M)$ called a **semi-ring.** To bring out the analogy with famiar algebraic structures, one can replace '$\cup$' by '+' on the right hand side of 15.7.

Proof. This result is well-known in the Formal Language literature [5, 61].

This semi-ring structure on $L(M)$ is the basic one for the Chomsky-Shutzenberger theory [8] of the recursive structure of languages . This theory essentially involves a sort of 'non-commutative algebraic geometry' based on the semirings $L(M)$, developed by analogy with the way classical algebraic geometry is defined in terms of the (commutative, associative) algebra of polynomials in 'commuting' variables over the complex number field.   What is most important from a qualitative and algebraic point of view is that, using this semi-ring structure, one can define Languages L, i. e. elements of $L(M)$, as solutions of algebraic equations.

Let us turn to the description of some categories and functors associated with monoids and languages.   Until further notice, we will make the following   assumption:

$$M \text{ is a free monoid generated by a subset } V. \qquad (15.9)$$

One reason for focussing on 15.9 is that it is the natural algebraic setting for an ubiquitous procedure in Computer Science: Start with an collection V of objects called **the alphabet** and let M be the free monoid generated by V, i.e the set of strings of elements of V, with the monoid operation just that of putting strings together, as described by the following formula::

$$\text{If } m = (v_1, \cdots, v_n) \text{ and } m' = (v_1', \cdots, v_p'), \text{ then}$$

$$mm' = (v_1, \cdots, v_n, v_1', \cdots, v_p'), \qquad (15.10)$$

$$\text{for } v_1, \cdots, v_n, v_1', \cdots, v_p' \; \varepsilon \; V$$

With the aid of this monoid product, we can write an m of form as follows:

$$m = v_1 \cdots v_n. \tag{15.11}$$

We can sum up what has been proved in the following way:

Theorem 15.2 . Any free monoid generated by V is isomorphic to the algebra of 'words' constructed from the 'alphabet' V, with the monoid product just that of juxtaposition of strings.

Example. The Language of the Propositional Calculus.

This is a standard topic in Logic, that is also important as a simple prototype of many computer languages. . Start off with a alphabet A consisting of the following objects:

$$\text{Elements of } P, \cap, \cup, \neg, (\,,\,) \tag{15.12}$$

P is a set; The elements of P are called the **atomic propositions.** $\cap$ and $\cup$ are called the **disjunction** and **conjunction** symbols. $\neg$ is the **negation** symbol,

Let A* be the strings in this alphabet. The Propositional Calculus is a Language based on the alphabet A i.e. a subset **P(A)**of A* defined by certain recursion rules, the so-called **well-formed formulas** of Logic. The following recursion rules determine the elements of**P(A)**:

$$\text{The elements of } P \text{ are well-formed formulas} \tag{15.13}$$

If F, F' are well-formed formulas, then so are:

$$(F \cap F'), (F \cup F') \text{ and } \neg(F) \tag{15.14}$$

P(A) is now defined as the smallest subset of A* satisfying 15.13-15.14.

Example. The Computer Language LISP

I will not attempt a definitive definition of LISP. The following properties seem to be characteristic:

LISP is generated as a monoid by a finite set, $\qquad$ (15.15)
whose elements consist of **atoms**,

and the symbols: (, $\cdot$, and).

If W and W' ε LISP, then (W$\cdot$W') ε LISP. $\qquad$ (15.16)

Here are some useful properties of the representation of the free monoid M in terms of its 'alphabet' V.

Theorem 15.3. For m ε M, set:

$$|m| = n = \text{length } m. \qquad (15.17)$$

and, for n ε **N**,

$$M_n = \{m \ \varepsilon \ M: |m| = n\} \qquad (15.18)$$

Then, for n, n' ε **N**,

$$M_n M_{n'} = M_{n+n'} \qquad (15.19)$$

Also, the map m --> |m| is a homomorphism from the monoid M to the monoid formed by the nonnegative integers **N** under addition. 15.19 means that the map

$$n \ \text{-->} \qquad M_n$$

defines a homomorphism of the semigroup of N under addition to the semigroup formed by the elemets of **L**(M) under multiplication.

Historically, the importance of monoids in Applied Mathematics has come from their role as the basic algebraic structures in Automata and Formal Language theory. Here is one way to construct an Automata (and a Feedback Control System) from a monoid action of M on X. Let

$$U \subset M \qquad\qquad (15.20)$$

be a subset of M that contains the identity element. We can then construct a 'discrete-time' automaton with input space U and state space X as follows:

The input sequences $u_0, \ldots, u_n$ are elements of U $\qquad (15.21)$

For an initial state x_0, the result of applying the 'machine' is the sequence $x_0, \ldots, x_{n+1}$ of elements of X defined as the solution of the following difference equation:

$$x_{j+1} = u_j x_j \qquad\qquad (15.22)$$
$$\text{for } j = 0, \ldots, n.$$

In dynamical systems terms,

$$j \dashrightarrow x_j \qquad\qquad (15.23)$$

is the **orbit** of the **flow** on X generated by the (open-loop) control 15.21.

Let us introduce some of the mathematical terminology of the Control literature: It differs somewhat from that in the Automata literature. The important feature of the Control point of view- especially for the mathematician - is that it is closely linked to the 'dynamical system' methodology that is being applied with great frequency and success to Computer Science; see the work of Blum, Shub and Smale [6].

Definition. Given the monoid M acting on the set X, and the subset U of M, the ordered triple:

$$(U, M, X) \tag{15.24}$$

is called a **(discrete-time) control system with inputs U and state space X.** A (possibly partial) map:

$$N \dashrightarrow U \times X$$

$$j \dashrightarrow (u_j, x_j) \tag{15.25}$$

is called an **open-loop trajectory** of the system if the difference equation 15.22 is satisfied.

We can now introduce the **feedback** or 'closed loop' concept in this algebraic framework.

Definition A **feedback** law for the system 15.22 is a map:

$$K: X \dashrightarrow U. \tag{15.26}$$

$$x \dashrightarrow K(x).$$

A trajectory $j \dashrightarrow (u_j, x_j)$ of the system (U, M, X) is said to be a **closed loop trajectory associated with the feedback law 15.26** if it satisfies the following conditions:

$$u_j = K(x_j) \tag{15.27}$$

The 'closed loop' condition 15.27 on the trajectory amounts to combining the first order system of difference equations 15.22 and the zero-th order system 15.27 to obtain a composite system whose solutions are the 'closed loop trajectories'. In 'language' terms, this amounts to restricting the language 'accepted' by the Automata to a 'sublanguage'. This point of view towards the language version of feedback control is a basic one in the work of Wonham and Ramadge. From the point of view of the theory of 'generalized differential systems' under development in this paper this amounts to finding a 'sub-system' of the 'differential system' defining the open-loop trajectories of the control system.

Let us now describe how this monoidal description of control systems is related to the usual one to be found in the Control literature:

Theorem 15.4.   Let X and U be sets, and let:

$$f: U \times X \rightarrow X \tag{15.28}$$
$$(x, u) \rightarrow f(u, x)$$

be a map.   Define a map:

$$\mathbf{f}: U \rightarrow \mathbf{MAP(X)} \tag{15.29}$$

by 'currying' the relation 15.28.   Explicitly, assign to each fixed $u \, \varepsilon \, U$ the map:

$$\mathbf{f}(u): x \rightarrow f(u, x) \tag{15.30}$$

of X into itself. **f** is then a map:

$$\mathbf{f}: U \rightarrow \mathbf{MAP(X)} \tag{15.31}$$

Let U^* be the free monoid generated by U, i..e the set of strings composed of elements of U.   Then, the map **f** can be extended to a monoid homomorphism:

$$\mathbf{f^*}: U^* \rightarrow \mathbf{MAP(X)} \tag{15.32}$$

which agrees with **f** on U. This defines a transformation action of the monoid U^* on X, i.e. a map

$$U^* \times X \rightarrow X \tag{15.33}$$

satisfying the conditions .

Now, consider the discrete-time control system associated with f:

$$x(n+1) = f(x(n), u(n)) \qquad (15.34)$$

The trajectories n --> x(n) of the control system, i.e. the solutions of the difference equation 15.34 for some input curve n --> u(n), may then be considered as orbits of the action of U^* on X. Thus, there is an essential equivalence at this set-theoretic level between 'control systems' as developed in the engineering and mathematics literature and 'automata' as developed in the Logic and Computer Science literature.

Proof. All of this follows directly from general set-theoretic principles.

Remark.   This identification of the language-theoretic and set-theoretic description of Automata is a basic link between Control Theory and Computer Science.

Theorem 15.5. Let M be a monoid which acts as tranformations

$$(m, x) --> mx \qquad (15.35)$$

on a set X. Let M' be a subset of M, and let M'^* be the free module generated by the elemets of M'. (M'^* consists of the 'words' constructed from elements of M' with the monoid product just concatenation of the 'letters' in the words.) Then, there is a unique monoid homomorphism:

$$M'^* --> M \qquad (15.36)$$

which is the identity map on the subset M' of M. In this way, the monoid M'^* acts on X. This action defines an 'automata' with X as state space.

Proof. The existence of the homomorphism   follows from general algebraic principles of 'freeness' or 'universality'.   The rest of the statements are obvious.

A basic concept in Automata theory is that of a 'language' being 'accepted' by an automata an a collection of 'final states'.   Here is one way of codifying this concept in algebraic terms:

Definition.   Let M', M'^*, M, and X be as defined in the statement of Theorem 15.5. Let L be a subset of M'^*, i.e. a 'language' . L is said to be **accepted** by a subset Y of X with respect to an initial state element $x_0 \ \varepsilon \ X$ if the following condition is satisfied:

$$Lx_0 \subset Y. \tag{15.37}$$

Remarks.   Condition 15.37can be interpreted algebraically.   The action of the monoid on X defines a mapping:

$$\pi: M'^* \dashrightarrow X \tag{15.38}$$

defined as follows:

$$\pi(m) = mx_0 \ \text{ for } m \ \varepsilon \ M'^* \tag{15.39}$$

15.37 means that:

$$L \text{ lies in the inverse image of } Y \text{ under } \pi. \tag{15.40}$$

In case X is finite, the 'combinatorics' of the situation restricts the algebraic structure of L via the 'pumping lemma' [37] of automata theory.   (It is based on the fact that, if $n+1$ identical things are put into n boxes, at least one box must have two things.)

A basic theorem of Kleene completely characterizes the possible algebraic structure of the language L if X is   finite.   Since most of the languages used in Computer Science do not satisfy Kleene's conditions, in order that they be 'accepted' the automata which does so must have an infinite state space X. The famous Turing Machines seem to be what Computer Scientists and Logicians consider as the simplest such automata.

Turing Machines have a state space with an interesting algebraic and geometric structure: It is a composite of a 'tape' T with an abelian monoid structure on which one can write 'symbols' from a finite set S. There are several monoids involved here. Start off with the following two monoids:

The free monoid constructed from the alphabet S   (15.42)

The monoid of translations of the abelian monoid T   (15.43)

Form the direct product monoid U:

$$U = S \times T. \tag{15.44}$$

Now, form the monoid:

$$\mathbf{L}(U) \tag{15.45}$$

of sublanguages of U. Turing Machines now deal with submonoids of $\mathbf{L}(U)$, and actions of these submonoids on products of the space of the 'tape' T and spaces X on which S acts.

16.   PROCESSES, PRESHEAVES AND TRAJECTORIES

We have seen that there are close relations between control systems, systems of ordinary differential or difference equations - classically, called 'Monge Systems' - classical recursive functions, von Neumann- style computer programs, etc.   It is very natural for a differential geometer to think about possible unifications and generalizations, in the style of the general theory of 'Differential Systems'.   Such a theory might also serve as a foundation for the computing scientist's dreams of a theory of 'parallel' or 'connectionist' computation.   I present here several ideas about what might go into such a general theory and describe some algebraic structures based on Category Theory which might serve as a setting for a general theory of Processes and Recursive Systems.

Let X be a set with a partial ordering

$$\{ (x, x'): x \leq x'\} \tag{16.1}$$

given between its elements.   As we have seen, this partial ordering defines a category, denoted as $\{ X, \leq\}$ or $\mathbf{X}$, with X as the set of objects, and with morphisms the ordered pairs (x, x') that satisfy 16.1.

Definition.   A **Process**, parameterized by the category $\{ X, \leq\}$, is a contravariant functor from $\{ X, \leq\}$ to the category **SET**.   Thus, such a Process, denoted by σ, assigns to each $x \in X$ a set $\sigma(x)$ and to each pair (x, x') satisfying 16.1 a map:

$$\sigma(x, \ x'): \sigma(x') \dashrightarrow \sigma(x). \tag{16.2}$$

The following version of the associative law is to be satisfied:

For each triple (x, x', x") $\in$ X x X x X such that $x \leq x'$ and $x' \leq x''$, we have:

$$\sigma(x, \ x'') = \sigma(x, \ x') \, \sigma(x', \ x''). \tag{16.3}$$

In addition, the following identity relation is satisfied:

$$\sigma(x,\ x) = \quad \{\text{identity map}\} \qquad (16.4)$$
$$\text{for each }\ x\ \varepsilon\ X.$$

We shall denote such a Process as follows:

$$\sigma = \{\sigma(x),\ \sigma(x, x'),\ x\ \varepsilon\ X,\ x \leq x' \} \qquad (16.5)$$

σ is a **presheaf** if the set X is a collection of the open subsets of a topological structure on another set , with the ordering on X just that induced by the relation of set-theoretic inclusion.

Remark. A next step in generality might be the case where the target category is a more general topos than **SET**. I will leave this possibility to later work.

Definition. Let $\sigma = \{\sigma(x), \sigma(x, x'), x \in X\}$ be a Process, as defined above. Let U be a subset of X. A map

$$\tau : U \rightarrow \mathbf{SET}, \qquad\qquad\qquad (16.6)$$
$$x \rightarrow \tau(x),$$

is a **trajectory** associated with σ if it satisfies the following conditions:

$$\tau(x) \in \sigma(x), \qquad\qquad\qquad (16.7)$$

$$\sigma(x, x')(\tau(x')) = \tau(x) \qquad\qquad\qquad (16.8)$$

for all $(x, x') \in U \times U$ such that $x \leq x'$. The set of all such trajectories defined on the subset U of X is denoted by

$$\tau(\sigma, U). \qquad\qquad\qquad (16.9)$$

The assignment

$$U \rightarrow \tau(\sigma, U) \qquad\qquad\qquad (16.10)$$

defines a Process. In fact, it is a presheaf, with source category the set of all subsets of X.

Remark. The interesting examples of Processes and presheaves are those defined by some algorithmic procedure analogous to a differential, integral, or algebraic equation. The Chomsky-Schutzenberger theory [8] of recursively-defined languages is an example of such an algorithmic procedure. Another example is the classical notion of 'recursive function'. Following Dana Scott's ideas [62-65, 67, 69], this can be put into a very general setting by defining appropriate topologies on the presheaf $\{\tau(\sigma, U)\}$, thereby defining the notion of continuity for functors between presheaves, and then defining the 'Recursive Processes' as the fixed points of continuous endo-functors. This method has already been used in special cases in many computing science contexts. Another method - proposed in [40] - is to classify processes in terms of the 'complexity' of the 'languages' generated by the trajectories of the process.

Example.   Functors whose target categories are retracts.

From the point of view of category and sheaf theory, the various branches of mathematics can be distinguished by the nature of the target categories of the processes and presheaves that arise in their study.   For example, differential geometry deals with presheaves whose target categories are modules over the ring of smooth functions on Euclidean spaces.   Similiarly,   algebraic geometry often works with modules over commutative Noetherian rings. Computing science - especially the branch called Formal Language Theory [19, 46, 48] - often deals with functors whose target objects are subsets of a given set Y and whose target morphisms are 'retract mappings'.   Let us then review the   definitions associated with the term 'retract'.

Definition.   Let Y be a set and let Z be a subset of Y. A mapping:

$$\rho: Y --> Z \qquad\qquad (16.11)$$

is said to be a **retract** of Z is the set of fixed points of ρ, i.e if the following condition is satisfied:

$$Z = \{y \, \varepsilon \, Y: \rho(y) = y\} \qquad\qquad (16.12)$$

Definition. Let $\{X, \leq\}$ be a partially ordered set, and let $C(X, \leq)$ be the category associated with this structure.   Let Y be a set and let σ be a contravariant functor (i.e. a process) whose source category is $C(X, \leq)$ and whose target category is a subcategory of the category of all subsets of Y. σ is said to be of **retract type** if the morphisms of the target category consist of retract mappings.

Theorem 16.1 A contravariant functor σ whose source category is $C(X, \leq)$ and whose target category is a subcategory of the category of all subsets of Y is of retract type if and only if σ assigns to each x ε X a subset $\sigma(x) \subset Y$, and to each pair (x, x') ε X x X such that:

$$x \leq x', \tag{16.13}$$

it assigns a retract mapping:

$$\rho(x', x): \sigma(x') \dashrightarrow \sigma(x), \tag{16.14}$$

with:

$$\sigma(x) \subset \sigma(x'). \tag{16.15}$$

For each triple (x, x', x'') ε X x X x X such that:

$$x \leq x' \text{ and } x' \leq x'' \tag{16.16}$$

the following condition is to be satisfied:

$$\rho(x'', x) = \rho(x', x')\rho(x'', x') \tag{16.17}$$

Proof. These conditions just make explicit those implicit in the above Definition. Condition 16.17 assures that σ is a contravariant functor.

Q.E.D.

We are interested in this type of contravariant functor because its trajectories have properties which are useful for modelling typical Discrete Event System and Computing Science Process behavior. The following result describes the trajectory property in this case.

Theorem 16.2 Let

$$\sigma : C(X, \leq) \to (\text{subsets of } Y)$$

be a contravariant functor of retract type, as described by Theorem 16.1. Let τ be a mapping: $X \to Y$. Then τ defines a trajectory of σ if and only if the following conditions are satisfied:

For $x \in T$, $\tau(x) \in \sigma(x)$, whenever $(x, x') \in X \times X$ with:

$$x \leq x' \tag{16.18}$$

we have:

$$\rho(x', x)(\tau(x')) = \tau(x) \tag{16.19}$$

Proof. Follows directly from the definitions.

In order to provide examples of this structure that occur often in computing science, let us suppose given the following structures on the set Y:

a) Y is a monoid $\hspace{3cm}$ (16.20)

b) A map:

$$N \to (\text{subsets of } Y) \tag{16.21}$$
$$n \to Y_n$$

such that:

$$Y_n Y_k \subset Y_{n+k}. \tag{16.22}$$

Such a structure is called a **gradation** of Y. For each integer n, set:

$$Y(n) = Y_0 \cup \cdots \cup Y_n \qquad (16.23)$$

We then have:

$$Y(0) \subset \cdots \subset Y(n) \subset \cdots$$

$$(16.24)$$

$$Y(n)Yk) \subset Y(n+k)$$

A family $\{Y(n)\}$ of subsets of Y satisfying 16.24 is called a **filtration** of the module structure on Y.

If, further, Y is a free monoid generated by Y_1, we can define retraction maps of Y . For each pair of integer (n, p) such that:

$$1 \leq p \leq n \qquad (16.25)$$

define a map:

$$\rho(n, p): Y(n) \longrightarrow Y(p) \qquad (16.26)$$

by the following formulas:

$$\rho(n, p)(v_1 \cdots v_m) = v_1 \cdots v_p. \qquad (16.27)$$

if $p \leq m$, and

$$\rho(n, p)(v_1 \cdots v_m) = v_1 \cdots v_m \qquad (16.28)$$

if $m < p$

Theorem 16.3. Let N be the ordered set of the integers, considered as a category.   Then the assignments

$$n \longrightarrow Y(n) \tag{16.29}$$

$$(n, p) \longrightarrow \rho(n, p) \tag{16.30}$$

define a contravariant functor of retract type whose source category is the category associated with the poset N and whose target category is a subcategory of the category of all subsets of Y .

Proof. Again, follows from the definitions.

Let us now turn to the concept of a 'cross-section', which is more general than that of 'trajectory'.

17. CROSS-SECTIONS OF PROCESSES WHOSE SOURCE CATEGORY IS A FAMILY OF SUBSETS. FIBER SPACES.

Let X be a set. Let **U** be a collection {U} of subsets of X. Consider **U** to be partially-ordered under inclusion and also the category associated with the partial ordering. An object of **U** is then an element U of **U** and a morphism is a pair (U, U') of elements of **U** such that:

$$U \subset U' \qquad (17.1)$$

Suppose that σ is a process mapping **U** to **SET**. σ assigns to each element $U \in$ **U** a set $\sigma(U)$ and to each pair (U, U') of elements of **U** that   satisfy 17.1   a map:

$$\sigma(U, U'): \sigma(U') \dashrightarrow \sigma(U). \qquad (17.2)$$

such that, whenever U, U' and U" are three elements of **U** satisfying:

$$U \subset U' \subset U'' \qquad (17.3)$$

we have:

$$\sigma(U, U'') = \sigma(U, U')\sigma(U', U'') \qquad (17.4)$$

We have defined 'trajectories' as certain mappings from the objects of **U** to points of the sets which are objects of the target category and which fit together with respect to the action of the target morphisms. Let us now consider another, more general, concept, a "cross-section" associated with a process with source category **U**. Set:

$$\Gamma(U, \sigma) = \text{set of maps } \gamma : U \dashrightarrow \sigma(U) \qquad (17.5)$$

Definition. A **cross-section of** σ, generically denoted as γ , assigns to each $U \in$ **U** a map:

$$\gamma(U): U \to \sigma(U) \tag{17.6}$$

such that:

$$\gamma(U) = \sigma(U, U')\gamma(U') \qquad \text{restricted to } U \tag{17.7}$$
$$\text{for all } U, U' \in \mathbf{U} \text{ such that } U \subset U'.$$

Let:

$$\Gamma(\mathbf{U}, \sigma) = \text{collection of cross-section maps} \tag{17.8}$$

Remarks. A 'trajectory' is just a special sort of 'cross-section', namely a collection of mappings of $U \to \sigma(U)$ that are constant. The term 'cross-section' arose in mathematics in connection with fiber bundle theory. Let us briefly review the description of processes generated in this way.

Example. The pre-sheaf of cross-sections of fiber spaces.

Let X and Y be sets, i.e. objects of **SET,** and let $\pi : Y \to X$ be a map. Within **SET**, the triple (π, Y, X) will be called a **fiber space** if $\pi(Y) = X$, i.e if π is an onto map.

Remark. Within other subcategories of **SET**, we might adopt further conditions to define the term 'fiber space'. For example, within the category **TOP** of topological spaces and continuous mappings, it would be appropriate to require further that π have the 'covering homotopy' property. Within the category **DIFF** of differentiable manifolds and smooth mappings, we might require further that π be a submersion, i.e. that the mapping induced by π on tangent vector bundles be onto.

Let **U** be the category of all subsets of X, partially-ordered by inclusion. If U is an object of **U**, a map:

$$\gamma : U \dashrightarrow Y \tag{17.9}$$

is said to be a **cross-section map (in the geometric sense) of the fiber space** (π, Y, X) if:

$$\pi\gamma = \text{the identity map of U.} \tag{17.10}$$

Let:

$$\Gamma(\pi, U) = \text{set of cross-section maps of the fibre space} \tag{17.11}$$
$$(\pi, Y, X) \text{ above the subset U of the base space X.}$$

Then, the mapping:

$$U \dashrightarrow \Gamma(\pi, U) \tag{17.12}$$

defines a pre-sheaf.

Theorem 17.1. With the above notations, let $\gamma : X \dashrightarrow Y$ be a cross-section map of the fiber space (π, Y, X) defined on the whole base space. Assign to each subset U of X the map $\gamma(U): U \dashrightarrow Y$ defined as follows:

$$\gamma(U) = \gamma \text{ restricted to U} \tag{17.13}$$

Then, the mapping $U \dashrightarrow \gamma(U)$ defines a cross-section of the pre-sheaf 17.11-17.12 . Conversely, every cross-section of this pre-sheaf arises in this way from a globally-defined cross-section of the fiber space (π, Y, X).

Proof. Follows readily from the definitions.

Example. Cross-sections of local-product fibre spaces

Let (π, Y, X) be a fiber space defined within **SET**, as above. Let U be a set of subsets of X, partially-ordered by inclusion and thus considered as a category.

Definition. Let F be a set. The fiber space (π, Y, X) is said to have a **local product structure, with fiber F, relative to the category** U if the following condition is satisfied:

For each object U of **U**, there is a mapping:

$$\alpha(U): \pi^{-1}(U) \dashrightarrow F \tag{17.14}$$

such that:

For each U, the map:

$$y \dashrightarrow (\pi(y), \alpha(U)(y)) \tag{17.15}$$

of $\pi^{-1}(U) \dashrightarrow U \times F$ is one-one and onto.

Theorem 17.2. Supppose that the fiber space (π, Y, X) has a local product structure, with fiber F, relative to the category U of subsets of X., i.e. conditions 17.14-17.15 are satisfied. Let:

$$G(F) = \text{group of one-one maps of F onto itself.} \tag{17.16}$$

Let U and U' be two objects of **U**, i.e. subsets of X, which intersect. Then, there is a map:

$$\beta(U, U'): U \cap U' \dashrightarrow G(F) \tag{17.17}$$

such that:

$$\alpha(U)(y) = \beta(U, U')(\pi(y))(\alpha(U')(y) \qquad (17.18)$$
$$\text{for all } y \, \varepsilon \, \pi^{-1}(U \cap U')$$

Proof. For $x \, \varepsilon \, U \cap U'$, the points on the fiber $\pi^{-1}(x)$ have two different 'coordinates' or 'labels' as points of F, via the maps $\alpha(U)$ and $\alpha(U')$. $\beta(U, U')(x)$ is the mapping of F into itself that compares one such F-labelling to the other. That $\beta(U, U')(x)$ is a one-one map of F onto itself follows from 17.15.

$$\textbf{Q.E.D.}$$

In fiber bundle theory, one goes on to study the mapping:

$$(U, U') \dashrightarrow \beta(U, U') \, \varepsilon \, G(F) \qquad (17.19)$$

It is a '1- cocycle with coefficients in the group G(F)'. We will not pursue this direction here.

Remarks. This is the essence of an argument that is used throughout differentilal geometry, particularly in differentiable manifold and fiber bundle theory. A terminology and intuition invented by Ehresmannn is very useful here. The maps $\alpha(U)$ are 'charts' of an 'atlas' parameterized by U.

The prototype of this sort of structure is the 'classical tensor analysis' that Einstein used in the construction of his Theory of General Relativity. In modern language, it can be described as follows. Let X be a differentiable manifold, (m, n) a pair of integers, and let (π, Y, X) be the fiber space over X defined as follows:

$$Y = \{(x, \tau): \tau \text{ belongs to the tensor product} \qquad (17.20)$$
$$\text{over the reals of m copies}$$
$$\text{of the tangent vector space}$$
$$X_x \text{ and n copies of the dual}$$
$$\text{vector space } X_x{}^*\}$$

$$\pi: Y \dashrightarrow X \text{ is defined by:} \qquad (17.21)$$
$$\pi(x, \tau) = x.$$

(π, Y, X) is called the **bundle of m-contravariant, n-covariant tensors associated with the manifold structure on X.** Local coordinate systems for this manifold structure define the mappings $\alpha : \to F$, where F is the vector space R^{dmn}, with d = dimension of X, and U is a coordinate neighborhood.

Example. The fiber spaces that appear in the theory of machine vision and signal processing

Our examples of Processes and presheaves have so far been taken from mathematics. Let us now briefly indicate how similiar structures can be associated with situations that occur very naturally in the theory of vision and signal processing.

Let X be a portion of the Euclidean plane R^2. For example, X could be part of a lattice in R^2, e.g. a set of points with integer coefficients. Let Y be a set, and let

$$Z = X \times Y \tag{17.22}$$

$$\pi: X \times Y \to X$$
$$\pi(x, y) = x$$

be a product fiber space, with fiber Y. Think of 'Y' as some sort of 'intensity space'. Let U be a subset of X. A geometric cross-section cross-section: $\gamma: U \to Z$ then defines a map: $U \to Y$. The value in Y assigned to a point x of X then represents a value of the 'intensity' attached to the 'picture' γ at the point x. Thus, there is a completely 'trivial' isomorphism between 'pictures' and ' geometric cross-sections'. Often, in the engineering applications one wants to somehow keep track of this representation when various transformations are applied to the 'picture'. For example, think of the 'picture' after some sort of 'transmission' or 'filtering' process is applied. It is possible that the more general algebraic notions of 'cross-section' explored in this Section might be useful in for this purpose.

Let us turn now to the study of the analogue of the concept of 'differential operator' and the resulting generalization of the 'Ehresmann jet spaces'.

18. OPERATORS ON CROSS-SECTIONS AND EHRESMANN SPACES

Definition. Let σ and σ' be Processes with the same source-category U, associated with family $\{U\}$ of subsets of X partially-ordered by inclusion. An **operator** between cross-sections of σ and σ' is defined as a collection

$$\{\Delta(U): U \; \varepsilon \; \text{objects of } U \} \tag{18.1}$$

of maps

$$\Delta(U): \Gamma(U, \sigma) \dashrightarrow \Gamma(U, \sigma') \tag{18.2}$$

satisfying the following compatibility conditions whenever U, U' are two elements of U such that $U \subset U'$:

$$\Delta(U)\sigma(U', U) = \sigma(U', U)\Delta(U'). \tag{18.3}$$

Theorem 18.1. Let $\mathbf{U}$ be the category associated with a family $\{U\}$ of subsets of X partially-ordered by inclusion. Let σ and σ' be processes whose source category is $\mathbf{U}$. Let Δ be an operator from cross-sections of the process σ to cross-sections of the process σ', satisfying condition 18.3. Let x be a point of X, and let U and U' be two elements of $\mathbf{U}$ such that:

$$x \in U \subset U'. \tag{18.4}$$

Suppose that γ and γ' are two element of $\Gamma(U, \sigma)$. Suppose that:

$$\Delta(U)(\gamma)(x) = \Delta(U)(\gamma')(x). \tag{18.5}$$

Then,

$$\Delta(U')(\gamma)(x) = \Delta(U')(\gamma')(x). \tag{18.6}$$

In other words, equality of the value at x of Δ on a cross-section of σ is independent of the representative chosen for the cross-section.

Proof. Follows from 18.3, 18.4 and 18.5

Theorem 18.2. Let us suppose that the category **U** defined above satisfies the following condition:

For each pair (U, U') of objects of **U** , and each point

$$x \, \varepsilon \, U \cap U'. \tag{18.7}$$

there is a third element U" ε Z such that:

$$x \, \varepsilon \, U", \text{ and } U" \subset U , U" \subset U' \tag{18.8}$$

Let σ and σ ' be processes with **U** as source category. Let Δ be an operator from cross-sections of the process σ to cross-sections of the process σ', satisfying condition18.3. Let x be a point of X, and let U and U' be two objects of **U** such that x ε U $\cap$ U'. Suppose that γ and γ ' are two element of $\Gamma(U, \sigma)$. Suppose that:

$$\Delta(U)(\gamma)(x) = \Delta(U)(\gamma ')(x). \tag{18.9}$$

Then,

$$\Delta(U')(\gamma)(x) = \Delta(U')(\gamma ')(x). \tag{18.10}$$

In other words, equality of the value at x of Δ on a cross-section of σ is independent of the representative chosen for the cross-section.

Proof. Follows from Theorem 17.1, applied to the inclusions U" $\subset$ U and U" $\subset$ U'.

Q. E. D.

Remark. In case conditions 18.7 and 18.8 are satisfied, it makes consistent sense to use the notation:

$$\Delta(\gamma)(x) = \Delta(\gamma ')(x) \tag{18.11}$$

to mean that the following condition is satisfied:

$$\Delta(U)(\gamma(U))(x) = \Delta(U)(\gamma'(U))(x) \tag{18.12}$$

for all $U \, \varepsilon$ objects of **U** such that $x \, \varepsilon \, U$.

These conditions enable us to give an invariant meaning to the idea that an operator **takes the same value at a point of X on a pair of cross-sections of a process.**

Definitions. These concepts enable us now to define the generalization of the Ehresmann jet spaces. Let Δ be a set of operators on the cross-section space of the proccess σ which satsifies 18.7-18.11. Introduce the following equivalence relation on the product space $X \times \Gamma(\sigma)$:

Two elements (x, γ), (x, γ') of $X \times \Gamma(\sigma)$ are

equivalent with respect to Δ

if the following conditions are satisfied:

$$x = x' \tag{18.13}$$

$$\Delta(\gamma)(x) = \Delta(\gamma')(x) \text{ for all } \Delta \, \varepsilon \, \Delta. \tag{18.14}$$

The quotient of $X \times \Gamma(\sigma)$ under this equivalence relation is called the **Ehresmann space associated with the presheaf** σ **and the set of operators** Δ. It is denoted as follows:

$$J^{\Delta}(\sigma) \tag{18.15}$$

The Cartesian projection map $X \times \Gamma(\sigma) \dashrightarrow X$ passes to the quotient with respect to this equivalence relation to define a map:

$$\pi^{\Delta}: J^{\Delta}(\sigma) \dashrightarrow X \tag{18.16}$$

called the **canonical projection**. For x ε X, let:

$$J^\Delta (\sigma)(x) = (\pi^\Delta)^{-1}(x)$$

be the fiber of the map π^Δ above the point x.

We can now define other operations of the Ehresmann theory. Here is the definition of the 'prolongation' concept.:

Definition. Let $\gamma \in \Gamma(\sigma)$. For $x \in X$, let:

$$j^\Delta(\gamma)(x) \in J^\Delta(\sigma)(x). \tag{18.18}$$

be the equivalence class of the equivalence relation 18.13 to which the pair (x, γ) belongs. As x varies, we obtain a map:

$$j^\Delta(\gamma): X \longrightarrow J^\Delta(\sigma) \tag{18.19}$$

called the **prolongation** or **jet** of the cross-section γ.

Now, we can define the basic concept which generalizes 'difference' or 'differential' equations and yet is broad enough to cover the 'recursive' situations which appear in Logic and Computing Science:

Definition. A subset $DE \subset J^\Delta(\sigma)$ is said to define a **Recursive System relative to Δ**. A presheaf cross-section $\gamma \in \Gamma(\sigma)$ is said to be a **solution** if the following condition is satisfied:

$$j^\Delta(\gamma)(X) \subset DE. \tag{18.20}$$

Let us turn to the generalization of the concept of a 'difference operator'.

Definition. Let σ and σ' continue as two processses with the same source category **U**. Assume that **U** satisfies conditions 18.7-18.11. Let Δ and Δ' be a collection of operators mapping the cross-sections of σ and σ' (respectively) into themselves. Let D be an operator sending cross-sections of σ into cross-sections of σ'. D is said to be an **Ehresmann operator relative to Δ and Δ'** if the following condition is satisfied.

For each $x \varepsilon X$, and each pair γ , $\gamma_1 \varepsilon \Gamma(U, \sigma)$ that satisfies
condition 17.22 at the point x, the images $D(\gamma)$, $D(\gamma_1)$ (18.21)
also satisfy 17.22 at x relative to Δ '. In other words,
D defines a morphism between the equivalence relations
on $X \times \Gamma(U, \sigma)$ and $X \times \Gamma(U, \sigma')$

Condition 17.29 then guarantees that D passes to the quotient to define a
map:

$$D(\Delta, \Delta'): J^\Delta(\sigma) \longrightarrow J^{\Delta'}(\sigma') \tag{18.22}$$

We have now extended the definitions of the basic concepts of
differential system theory, as developed using the Ehresmann-Spencer-
Goldschmidt point of view. We now turn to describing a more special
algebraic method for defining such operators, patterned after the classical
theory of 'difference equations'.

Example. Difference operators and equations associated with transformation semigroups

Let us now use semigroups acting on a set X to define the operators Δ associated with Ehresmann spaces. For each positive integer r let X^r be the Cartesian product of copies of X. Let S be a semigroup, with the binary operation denoted multiplicatively:

$$(s_1, s_2) \longrightarrow s_1 s_2$$

S^r can be made into a semigroup, using the following multiplication rule:

$$(s_1, \ldots, s_r)(s_1', \ldots, s_r') = (s_1 s_1', \ldots, s_r s_r') \tag{18.23}$$

Let S act as a transformation semigroup on the set X:

$$(s, x) \longrightarrow sx. \tag{18.24}$$

Then, the semigroup S^r - with multiplication defined by 18.23 - can be made to act on X^r according to the following formula:

$$(s_1, \ldots, s_r)(x_1, \ldots, x_r) = (s_1 x_1, \ldots, s_r x_r) \tag{18.25}$$
$$\text{for } (s_1, \ldots, s_r) \, \varepsilon \, S^r, \, (x_1, \ldots, x_r) \, \varepsilon \, X^r$$

Let U be a subset of X such that:

$$s_i(U) \subset U \quad \text{for } i = 1, \ldots, r. \tag{18.26}$$

Let Z and Y be sets and let:

$$\gamma: U \longrightarrow Y \tag{18.27}$$

$$F: Y^r \longrightarrow Z \tag{18.28}$$

be maps. Set:

$$\Delta(\gamma)(x) = F(\gamma(s_1 x), \ \ldots, \gamma(s_r x))$$ (18.29)

Let us sum up as follows:

Theorem 18.3. With the above notations, and with 18.26 satisfied, let:

$$\Gamma(U, Y) = \text{set of maps: } U \dashrightarrow Y.$$ (18.30)

Define Δ by formula 18.29. It maps:

$$\Gamma(U, Y) \dashrightarrow \Gamma(U, Z).$$ (18.31)

and is called the **difference operator** associated with the
r-tuple **s** and the map F.

Proof.   Condition 18.26 shows that formula 18.29 defines a map as
indicated in 18.31.

Q. E. D.

By specializing the choice of S acting on X, we can obtain, using 18.30,
the difference operators considered in Section 11. For example, suppose that:

$$X = N;$$ (18.32)

$$S \text{ is the addditive semigroup of } N$$ (18.33)
$$\text{acting on itself.}$$

Formula 18.30 then provides a system of ordinary difference equations, of
the type considered above.

Here is another choice:

$$X = N \times A; \quad A \text{ is a set of 'parameters'.} \qquad (18.34)$$

S is the addditive semigroup of N
acting on X as follows:

$$s(n, a) = (n + s, a) \text{ for } n \, \varepsilon \, N, \, s \, \varepsilon \, S, \, a \, \varepsilon \, A. \qquad (18.35)$$

18.29 then takes the following form:

$$\Delta(\gamma)(n, a) = F(\gamma(n + n_1, a), ..., \gamma(n + n_r, a)) \qquad (18.36)$$

for

$$s = (n_1, ..., n_r)$$

Many familiar systems can be obtained by specialization of 18.36. For example, A could be a set of 'input' or 'control' parameters, as in control theory, or a set of 'recursion parameters', as in classical recursive function theory .

19. LINEAR DIFFERENTIAL OPERATORS

The game-plan of this work is to delineate some basic mathematical objects associated with what I see as 'geometric computing science', in analogy with the basic concepts of manifold-based differential geometry. As preparation for the definition of a class of linear difference operators to be given in Section 20, in this Section I will briefly review without proofs the definitions associated with the concept of a 'linear differential operator'. This concept is a fundamental one for mathematical physics as well.

We deal here only with operators that map vector-valued functions on a given manifold to another set of functions on the same manifold. Let M be a finite dimensional, infinitely differentiable manifold.   Set:

$\mathbf{O}(M)$ = the poset of subsets of M that are open   (19.1)
        in the manifold topology, considered
         also as a category via the partial ordering
        induced by inclusion.

$\mathbf{FO}(M)$ =   the category whose objects are the algebra
        (over the real numbers) of   (19.2)
         real-valued, infinitely differentiable
       functions on open subsets of M and whose
         morphisms are the restriction maps obtained when
one open subset of M is contained in another

Remark. The map which assigns to an object O of $\mathbf{O}(M)$, i.e. each open subset of M, the algebra F(O) of real-valued, infinitely differentiable functions on it, is a contravariant functor, i.e. a pre-sheaf. This pre-sheaf is the basic algebraic structure of differential geometry. Similiarly, the basic algebraic structure of computing science is the functor:

$$X \dashrightarrow X^* \qquad\qquad (19.3)$$

which assigns to a set X the monoid X^* of strings consisting of elements of X. As we shall see, the theory of linear differential operators can be formulated in terms of the pre-sheaf 19.2.   Similarly, much of computing science involves the various algebraic structures definable in terms of the functor 19.3.

Let M be a manifold. Let O be an object of $\mathbf{O}(M)$, i.e an open subset of M. Let W and W' be real vector spaces. Form:

$$L(W, W') = \text{real vector space of linear maps: } W \dashrightarrow W', \qquad (19.4)$$

and the following tensor product:

$$F(O) \otimes W \qquad (19.5)$$

The symbol '$\otimes$' in 19.5 denotes tensor product over the field of real numbers of real vector spaces. The real vector space 19.5 may also be considered as a module ove the ring $F(O)$. This module structure is defined as follows:

$$f(f' \otimes w) = (ff') \otimes w \text{ for } f, f' \varepsilon F(O), w \varepsilon W. \qquad (19.6)$$

Theorem 19.1. The elements of the real vector space $F(O) \otimes W$ may be identified with the vector space of smooth maps: $O \dashrightarrow W$. This isomorphism assigns to each $f \otimes w$, $f \varepsilon F(O)$, $w \varepsilon W$, the following map:

$$p \dashrightarrow f(p)w$$

of $O \dashrightarrow W$.

Definition. Let:

$$\mathbf{FO(M)} \otimes W \qquad (19.8)$$

be the category whose objects are the real vector spaces

$$F(O) \otimes W \qquad (19.9)$$

and whose morphisms are the restriction maps obtained when one open subset of M is contained in another. It is called the **category of vector-valued smooth maps locally defined on M.**

For each object O of $\mathbf{O}(M)$, let $V(O)$ be the set of smooth vector fields on O. $V(O)$ has various algebraic and geometric structures which are basic to differential geometry. Each $V \varepsilon V(O)$ defines a smooth cross-section

$$V: O \longrightarrow T(M), \quad p \longrightarrow V(p) \qquad (19.10)$$

of the tangent vector bundle to O. Each $V \in V(O)$ defines a linear map:

$$F(O) \longrightarrow F(O), \quad f \longrightarrow V(f) \qquad (19.11)$$

called **Lie derivative with respect to the vector field V**. The operation 19.11 is a derivation of the commutative, associative algebra structure on F(O).

Let r be a non-negative integer. Let us now define the concept of 'linear differential operator of order r' in a purely algebraic way, by induction on r:

Definition. Let W and W' be real vector spaces, and let:

$$D: F(O) \otimes W \longrightarrow F(O) \otimes W' \qquad (19.12)$$

be a map which is R-linear with respect to the vector space structures. Then, D is said to be a **zero-th order linear differential operator** iff:

    D is F(O)-linear with respect to the F(O)-module   (19.13)
    structure defined by 19.6.

D is said to be an **r-th order linear differential operator** iff:

    For each $f \in F(O)$, the operator:

$$\sigma(f, D): f' \otimes w \longrightarrow D(ff' \otimes w) - fD(f' \otimes w) \qquad (19.14)$$

    is an (r-1)-st order linear differential operator.

This coordinate - independent definition can also be specialized to give the traditional representation of linear differential operators. We will do this in the manner pioneered by Elie Cartan, using his concept of 'moving frame':

Definition. A **moving frame** for an open subset O of M is an ordered set

$$(V_1, ..., V_m) \qquad (19.15)$$

of elements of V(O) such that:

$$\text{For each point } p \ \varepsilon \ O, \text{ the elements}$$
$$(V_1(p), ..., V_m(p)) \text{ form a basis for } M_p, \qquad (19.16)$$
$$\text{the tangent vector space to M at p.}$$

We can now describe a representation of linear differential operators for open subsets O of M in terms of these moving frames:

Theorem 19.2. Let O be an open subset of M which carries a moving frame $(V_1, ..., V_m)$ of vector fields. Let r be a non-negative integer. Let W and W' be real vector spaces. Let:

$$D: F(O) \otimes W \dashrightarrow F(O) \otimes W' \qquad (19.17)$$

be a linear differential operator of order r on O associated with the pair (W, W'), as defined above. Then, if:

$$r = 0, \qquad (19.18)$$

there is a smooth map:

$$A: O \dashrightarrow L(W, W') \qquad (19.19)$$
$$p \dashrightarrow A(p)$$

such that:

$$D(f \otimes w)(p) = f(p)A(p)(w). \qquad (19.20)$$

If $r = 1$, there is an $(m+1)$- tuple $(A, A^1, .., A^m)$ of maps of the form 19.19 such that:

$$D(f \otimes w)(p) = \sum_{1 \leq i \leq m} V_i(f)(p)A^i(p)(w) + f(p)A(p)(w) \qquad (19.21)$$

Similiar expansions of linear differential operators in terms of moving frames can be shown to hold for larger values of r as well.

These definitions have been 'local', associated with the open subsets defining the topology on M underlying the manifold structure. Let us now formulate the corresponding 'global' concept using category ideas.

Theorem 19.3. Let

$$D: \ F(O) \otimes W \text{--> } F(O) \otimes W' \qquad (19.22)$$

be a linear differential operator on an open subset of M. Let O' be an open subset of O. Then, there is a 'restriction' mapping assigning to D a linear differential operator

$$D': \ F(O') \otimes W \text{--> } F(O') \otimes W' \qquad (19.23)$$

such that:

$$D'(f \otimes w)(p') = D(f \otimes w)(p')$$
$$(19.24)$$
$$\text{for } f \ \varepsilon \ F(O), \ w \ \varepsilon \ W, \ p' \ \varepsilon \ O'.,$$
$$\text{where f' is the restriction}$$
$$\text{of f to } O'.$$

In words, 19.24 means that:

$$D'(f \otimes w) \text{ is the restriction to } O' \text{ of } D(f \otimes w) \qquad (19.25)$$

Definition. A linear differential operator associated with the manifold structure on M is a map associating with each open subset of M a linear differential operator:

$$D(O): F(O) \otimes W \dashrightarrow F(O) \otimes W' \qquad (19.26)$$

such that:

Whenever O, O' are open subsets of M such that: $O \subset O'$, we have:

$$D(O) = D(O') \text{ restricted to } O, \text{ in the sense} \qquad (19.27)$$
$$\text{that 19.25 is satisfied}$$

20. A CLASS OF RECURSIVE LINEAR OPERATORS

Let us now define another class of linear operators by analogy with the definition of linear differential operators given above . This class is general enough to cover certain types of linear difference and recursion operators.

Let **ABEL** be the category whose objects are the abelian groups and whose morphisms are the homomorphisms in the usual algebraic sense between abelian groups. Let X be a set, i.e. an object of the category **SET**, and let W be an abelian groups, i.e. an object of **ABEL**. Let:

$$\Gamma(X, W) = \text{set of maps } \gamma : X \dashrightarrow W \qquad (20.1)$$

$\Gamma(X, W)$ can be made into an abelian group by adding maps point-wise.

Theorem 20.1. The map:

$$X \times W \dashrightarrow \Gamma(X, W) \qquad (20.2)$$

indicated in 20.1.defines a functor from the product category **SET** x **ABEL** to the category **ABEL**.

Proof. 20.2 defines the map on 'objects'. It is left to the reader to show how the map on morphisms is defined.

Q.E.D.

In this Section we will define a class of maps:

$$D: \Gamma(X, W) \dashrightarrow \Gamma(X', W') \qquad (20.3)$$

that we will call **linear** in the sense that they are homomorphisms relative to the abelian group structure on their domains and range. To define this class, let X' be another set and let:

$$\alpha_1 : X' \dashrightarrow X \qquad (20.4)$$

$$\alpha_2 : X' \dashrightarrow X$$

$$\cdots$$

$$\alpha_n : X' \dashrightarrow X$$

$$\pi : X' \dashrightarrow X \qquad (20.5)$$

be given maps with domains and ranges as indicated in 20.4-20.5. Set:

$$L(W, W') = \text{the abelian group of} \qquad (20.6)$$
$$\text{homomorphisms from } W \text{ to } W'$$

Let

$$A_1, A_2, ..., A_n : X \to L(W, W') \tag{20.7}$$

be an n-tuple of maps with the indicated domain and range:

Given the above data, for $\gamma \in \Gamma(X, W)$, set:

$$D(\gamma)(x') = A_1(\pi(x'))\gamma(\alpha_1(x')) + \cdots + A_n(\pi(x'))\gamma(\alpha_n(x')) \tag{20.8}$$

We have now proved the following:

Theorem 20.1. Formula 20.8 associates with the data (20.4-20.7) a linear map D with domain and range indicated in 20.3.

Example. The basic difference operator.

Let X be a given set, and let W be an abelian group. Let us define a map:

$$\delta_1: \Gamma(X, W) \to \Gamma(X \times X, W) \tag{20.9}$$

by the following formula:

$$\delta_1(\gamma)(x, x_1) = \gamma(x) - \gamma(x_1) \tag{20.10}$$
$$\text{for } (x, x_1) \in X \times X$$

It has the following property:

$$\delta_1(\gamma) = 0 \text{ iff. } \gamma \text{ is constant.} \tag{20.11}$$

In diagrammatic terms, 20.11 means that:

The sequence
$$\delta_0 \qquad \delta_1$$

$$W \longrightarrow \Gamma(X, W) \longrightarrow \Gamma(X \times X, W) \qquad (20.12)$$

of linear maps is exact, i.e:

$$\text{image}(\delta_0) = \text{kernel}(\delta 1) \qquad (20.13)$$

where δ_0 is the map which assigns to each $w \ \varepsilon W$
the constant map whose value is w.

The formulation of this property in homological-algebra terms is no 'accident': In fact, the sequence 20.12 is just the beginning of the operators of the **Alexander-Kolmogoroff-Spanier cohomology theory** [66].

Example. Linear Difference Operators in the Classical Sense.

Let us now specialize to the following situation::

$$X = X'. \quad \pi = \text{identity map} \qquad (20.14)$$

$$\begin{array}{l} X \text{ is an abelian group, with the group operation} \qquad (20.15) \\ \qquad \text{denoted additively} \end{array}$$

$$\alpha_i(x) = x - h_i \text{ for } 1 \le i \le n, \qquad (20.16)$$

where $h_1, .., h_n$ are given elements of X.

($h_1, .., h_n$ play the role that is classically assigned to the **step-size** in difference schemes.)

Then, we have:

$$D(\gamma)(x) = A_1(x)\gamma(x - h_1) + \cdots + A_n(x)\gamma(x - h_n) \qquad (20.17)$$

which is, indeed, a linear difference operator in the classical sense.

FINAL REMARKS

In this paper, I have begun the job of preparing the ground for the synthesis of the ideas of differential geometry and computing science. I hope that the reader will begin see how it can serve as a unified mathematical formalism for reasoning about such more concrete subjects as the numerical analysis of ordinary differential equations, the links between Control and Automata Theory, the Theory of Processes in the sense of the Computing Scientists and the Theory of Discrete Event Systems in the sense of the Control Engineer.

BIBLIOGRAPHY

1. G. Ammar and C. Martin, The geometry of matrix eigenvalue methds, *Acta App. Math.*, 5, 239-278.

2. B. Anderson and R. Moore, *Linear Optimal Control*, Prentice-Hall, 1971.

3. M. Barr and C. Wells, *Category Theory and Computing Science*, Springer-Verlag, New York, 1990

4. J. Berstel and C. Reutenhauer, *Rational Series and Their Languages*, Springer- Verlag, 1984.

5. L. Blum, M. Shub and S. Smale, On a theory of computation and complexity over the real numbers: NP completeness, recursive functions and universal machines, *Bull. Amer. Math. Soc.*, 21 (1989), 1-46.

6. P. Caines, *Linear Stochastic Systems*, J. Wiley, 1988

7. E. Cartan, *Les Systemes Differentielles Exterieures et leurs Applications Geometriques*, Hermann, Paris, 1946.

8. N. Chomsky and M. P. Schutzenberger, The algebraic theory of context-free languages, in *"Computer Programming and Formal Systems"*, P. Braffort and D. Hirschberg, eds., North-Holland, Amsterdam, 1963

9. C. Ehresmann, *Oeuvres Completes*, Pts. 1 - 1 and 1 -2, Amiens, 1984.

10. C. Ehresmann, *Categories et Structures*, Dunod, Paris, 1965

11. S. Eilenberg, *Automata, Languages and Machines*, Vols. A and B, Academic Press, 1974.

12. S. Eilenberg and C. Elgot, *Recursiveness*, Academic Press, 1970.

13. M. Fliess, Matrices de Hankel, *J. Maths Pures Appl.* 53, 197-222

14. M. Fliess, Functionelles causales non lineaires et indeterminee non commutatives, *Bull. Soc. Math. France*, 109, 3-40

15. C. W. Gear, *Numerical Initial value Problems in Ordinary Differential Equations*, Prentice-Hall, 1971.

16. R. Goldblatt, *Topoi*, 2nd ed., N. Holland-Elsevier, 1984

17.   H. Goldschmidt, Existence theorems for analytic linear partial differential equations, *Ann. of Math.* **86**, 246-270 (1967).

18. E. Goursat, *Lecons sur le Probleme de Pfaff*, Hermann, Paris, 1922.

19.   M. A. Harrison, *Introduction to Formal Language Theory*, Addison-Wesley, Reading, Mass, 1978

20. W. S. Hatcher, *The Logical Foundations of Mathematics*, Pergamon Press, 1982

21.   M. Hennessy, *Algebraic Theory of Processes*, MIT Press, 1988

22. P. Henrici, *Discrete Variable Methods in Ordinary Differential Equations*, J. Wiley, 1962.

23. R. Hermann, A 'geometric' view of the dynamics of the trajectories of computer programs, *Acta App. Math.*, **18**, 145-182, 1990

24. R. Hermann, *Topics in the Geometric Theory of Linear Systems*, Math Sci Press, Brookline, MA, 1984.

25. R. Hermann, E. Cartan's geometric theory of partial differential equations, *Adv. Math.* **1**, 265-317 (1965)

26. R. Hermann, Geometric construction and properties of some families of solutions of nonlinear partial differential equations, *J. Math. Physics*, 24, 1983, 510-521

27. R. Hermann, On the accessibility problem of control theory, in J. Lasalle and S. Lefschetz (eds.), *Proc. Symp. on Diff. Eq.*, Academic Press, 1961.

28. R. Hermann, The Geometry and Lie Theory of the Matrix Riccati Equation, in *"Interdisciplinary Mathematics"*, vol. 20, 1979

29. R. Hermann, A Relation between the LR and QR Algorithms of Matrix Numerical Analysis and Lie Theory, in *"Interdisciplinary Mathematics"*, vol. 20, 1979

30. R. Hermann, Geometric aspects of potential theory in symmetric spaces, Part 3, *Math. Ann.*, 153 (1964), 384-394

31. R. Hermann, **Papers on Deformation Theory in Physics.in the 1960's.** Compactifications of Homogeneous Spaces and Contractions of Lie Groups, *Proc. Nat. Acad. Sci.* 51 (1964), pp. 456-461. The Gell-Mann Formula For Representations of Semisimple Groups, *Comm. Math. Phys.* 2 (1966), pp. 155-164.*Lie Groups for Physicists*, W. A. Benjamin Inc., 1966. Analytic Continuation of Group Representation, Part I, *Comm. Math. Phys.* 2 (1966), pp. 251-270; Analytic Continuation of Group Representation, Part II, *Comm. Math. Phys.* 3 (1966), pp. 53-74. Analytic Continuation of Group Representation, Part III *Comm. Math. Phys.* 3 (2966), pp. 75-97. Analytic Continuation of Group Representations, Part IV, *Comm. Math. Phys*, 5 (1967), pp. 131-156. Analytic Continuation of Group Representations, Part V, *Comm. Math. Phys*, 5 (1967), pp. 157-160. Analytic Continuation of Group Representations, *Comm. Math. Phys*, Part VI 6 (1967), pp. 205-225.

32. R. Hermann and A. Krener, Nonlinear controllability and observability, *IEEE Trans. Aut. Contr.*, **AC-22**, 728-740 (1977).

33. R. Hermann and C. Martin, Applications of algebraic geometry to system theory: The McMillan Degree and Kronecker Indicesof transfer functions as topological and holomorphic system invariants, *SIAM J. Control and Optimization*, 16 (1978), 743-755.

34. R. Hermann and C. Martin, Applications of algebraic geometry to systems theory, *IEEE Trans. Aut. Con.*, AC-22, 1877, 19-25.

35. R. Hermann and C. Martin, Lie and Morse Theory for Periodic Orbits of Matrix Riccati Equations, *Math System Theory*, 15, 277-284, 1982

36. R. Hermann and C. Martin, Feedback and pole placement for linear Hamiltonian systems, *Proc. IEEE*, 65 (1977), 841-848

37 J. Hilgert, K. H. Hoffman and J. D. Lawson, *Lie Groups, Convex Cones and Semigroups*, Oxford University Press, 1989

38. C. A. R. Hoare, *Communicating Sequential Processes*, Prentice Hall, 1985

39. J. Hopcroft and J. Ullman, *Introduction to Automata Theory, Languages and Computation*, Addison-Wesley, 1979

40. K. Inan and P. Varaiya, Finitely recursive process models for discrete event systems, *IEEE Trans Aut. Con.*,33, 626-638

41. A. Isidori, *Nonlinear Control Systems: An Introduction*, Springer-Verlag, Berlin, 1985.

42. M. Janet, *Lecons sur les Systems d'Equations aux Derivees*, Gauthier-Villars, Paris, 1929.

43. P. T. Johnstone, *Stone Spaces*, Cambridge University Press, 1982

44. R. E. Kalman, P. L. Falb, and M. A. Arbib, *Topics in Mathematical System Theory*, Mc Graw-Hill, 1969.

45. S. J. Kleene, Representation of events in nerve nets and finite automata, in "*Automata Studies*", C. Shannon and J. McCarthy eds., Princeton Univ. Press, 1956

46. W. Kuich and A. Salomaa, *Semirings, Automata, Languages.* Springer-Verlag, 1986

47. A. Kumpera and D. C. Spencer, *Lie Equations*, Princeton Univ. Press, 1972.

48. G. Lallement, *Semigroups and Combinatorial Applications*, Wiley, 1   978

49. J. Lambek and P. J. Scott, *Introduction to Higher-Order Categorical Logic*, Cambridge University Press, 1986.

50. S. MacLane, *Categories for the Working Mathematician*, Springer-Verlag, 1976

51. E. Manes, *Algebraic Theories*, Springer-Verlag, 1976.

52. E. G. Manes amd M. A. Arbib, *Algebraic Approaches to Program Semantics*, Springer-Verlag, New York, 1986

53. R. Milner, *Communication and Concurrency*, Prentice-Hall, 1989

54. R. Moll, M. Arbib amd A. Kfoury, *An Introduction to Formal Language Theory*, Springer-Verlag, 1988

55. D. Mumford, *The Red Book of Varieties and Schemes*, Springer-Verlag, 1988

56. J. F. Pommaret, *Systems of Partial Differential Equations and Lie Pseudogroups*, Gordon and Breach, New York, 1978

57 . P. Ramadge and M. Wonham, Supervisory control of a class of discrete event processes, *SIAM J. Control Opt.*, 25, pp. 206-230, 1987

58. G. Reyes, From Sheaves to Logic, in *Studies in Algebraic Logic*, ed. by A. Diagneault, MAA Studies in Mathematics, Vol. 9, 1974

59. C. Riquier, *Les Systemes d'Equations aux Derivees Partielles*, Gauthier-Villars, Paris, 1910

60. H. Rogers, *Theory of Recursive Functions and Effective Computability*, Mc Graw-Hill, New York, 1967

61. A. Salomaa and M. Soittola, *Automata-Theoretic Aspects of Formal Power Series*, Springer-Verlag, 1978

62. D. S. Scott, Some ordered sets in computer science, in "*Ordered Sets*", I. Rival, Ed., Reidel, Dordrecht, 1982, pp. 677-718.

63. D. S. Scott, Data types as lattices, *SIAM J. Comp.*, 5, 522-587, 1976

64. D. S. Scott, Logic and programming languages, *Comm. ACM*, 20, 634-641, 1977

65. D. S. Scott, Lectures on a mathematical theory of computation, in *"Theoretical Foundations of Programming Methodology"*, M. Broy and G. Schmidt, Eds., Reidel, 1982.

66. E. Spanier, *Algebraic Topology*, Mc Graw-Hill, 1966.

67. D. C. Spencer, Overdetermined systems of linear partial differential equations, *Bull. Amer. Math. Soc.* **75**, 179-239 (1969).

68. J. E. Stoy, *Denotational Semantics: The Scott-Strachey Approach to Programming Languages*, MIT Press, 1977

69. J. Thomas, *Differential Systems*, AMS Colloquium Publications, v. 21, New York, 1937.

70. S. Vickers, *Topology Via Logic*, Cambridge University Press, 1989

Acta Applicandae Mathematicae **18**: 145–182, 1990.
© 1990 *Kluwer Academic Publishers. Printed in the Netherlands.*

A 'Geometric' View of the Dynamics of Trajectories of Computer Programs

ROBERT HERMANN*
53 Jordan Road, Brookline, MA 02146, U.S.A.

(Received: 17 October 1988; in final form: 19 December 1989)

Abstract. This paper applies mathematical ideas which arise from differential geometry and control theory to computer programming theory. For the sake of simplicity, most questions of logic and language are put to the side. A computer program is regarded as a collection of data, considered as a collection of cross-sections of a fiber space whose base is a discrete space. These distinguished cross-sections are called the trajectories of the computer program system. For the simplest von Neumann type of program, the base space is the nonnegative integers: The trajectories may then often be thought of as the orbits of a discrete-time dynamical system. Feedback control theory enters when one attempts to provide a 'rational' way of generating such dynamical systems, satisfying certain conditions, such as termination and correctness. The greatest common divisor and bubble sort programs – standard examples in computer science textbooks – are treated as models for the development of a general theory.

Geometrically, we will also consider them as the discrete analogue of what are called path systems in differential geometry. These examples can also be treated as the analogue of closed-loop feedback control systems – themselves a sort of path system – that are 'driven' to the final answer by a process that is similar to the way that Lyapunov theory is used in the linear regulator problem of control theory. This use of the Lyapunov ideas of dynamical system theory gives a technique for proving the correctness and termination of some programs. In order to provide a mathematical structure to a programming theory that is, in some sense, analogous to the group-theoretic structure of fiber bundle differential geometry, an algebraic structure of a category which acts on the fiber space is associated with such programs. The computation of the binomial coefficients will also be treated as an example of a system of partial difference equations, leading to the description of some features of concurrent programs. Prolongations and correspondences are defined for programs, based on the analogous ideas for differential systems. Finally, some of the geometric background of modal logic is developed, particularly as it is used in following the temporal development of computer programs. The approach is related to the classical geometric notion of correspondence, and to the Jonsson–Tarski theory of operators on Boolean algebras.

AMS subject classifications (1980). 03G30, 06B30, 03B35, 6802, 68G15, 68C05, 9302, 93C55, 81D15.

Key words. Feedback, control systems, computer algorithms, path geometry, prolongations, categories.

1. Introduction

One of the main scientific problems of our time is the development of the theory of the mathematical nature and structure of computer programs. In this paper, I initiate a research program aiming towards the goal of understanding the mathematical nature

*This work was supported by grants from the Ames Research Center of NASA and the Applied Mathematics Program of the National Science Foundation.

of the computational results provided by programs rather than the linguistic or logical description. I particularly emphasize analogies with other parts of pure and applied mathematics, differential geometry and control theory.

Part of the difficulty in developing a precise mathematical structure for the theory of programming has been the complication of integrating language, logic, and computation. In this beginning effort, in order to simplify, I will mostly put to the side this aspect of programming theory, working in the conventional set-theoretic context of the mathematician. As partial justification for this simplification, I can cite the discipline of program semantics, one of whose aims is precisely to automate the translation from languages to mathematics. This has been treated extensively with powerful mathematical tools in several recent treatises [9, 24] to which the reader can refer. This paper will mainly deal with mathematical models that reflect the conventional von Neumann programming style, although I will provide a more general formulation: Indeed, this more general formulation might be useful in understanding the emerging world of concurrent and parallel formalisms.

Conventionally, a computer programming theory [1, 2, 3, 6, 9, 19, 20, 22, 23, 24] consists of two components: First, a *language*, i.e. a set of symbols and rules for combining them, and a *syntax* that picks out admissible collections of these symbols. Second, a *semantics* which assigns mathematical objects, such as sets and mappings, to syntactically admissible collections of symbols. Thus, 'semantics' acts much like a *functor* from categories that are 'linguistically and syntactically' defined to the set-theoretic categories of the 'working' pure or applied mathematician. (Note the recent book by Manes and Arbib [19], which makes explicit this connection between computer program semantics and category theory.)

This over-arching view of the nature of computer programming theory is essentially the world-view of the logician. There is a gulf between such a general picture and what must be done in practice to write a program for some computational purpose: It is basically the difference between pure and applied mathematics. My aim in this paper is to attempt to build a bridge between these two worlds, by defining, and beginning the job of analyzing the more concrete mathematical structures which seem to appear with great frequency on the applied side of programming theory.

Following an analogy suggested by me by Gerald Sussman, think of a program as the analogue of a differential equation. Then think of the computation produced as a result of the program as the analogue of the solution or trajectory of the differential equation. It is the collection of such 'trajectories' that will be studied in this paper, thus suggesting the idea that there is a dynamics (and a geometry) associated with a computer program.

The scientific analogies we are looking for extend beyond 'dynamics', The system of trajectories is not given a priori, but must be prescribed by the programmer to perform given tasks. Think of mechanics, which has rules (traditionally called rational mechanics) for translating the physical and kinematic data into dynamical equations. Unfortunately, there seems to be nothing like a Lagrangian for computer programming theory which will generate the equations for trajectories in some automatic way.

Instead, we shall look to feedback control theory for ideas about generating the desired trajectories. We take the view of the 'engineer' rather than the 'physicist' in thinking about computer programming: The trajectories are to be defined by some process analogous to feedback, subject to such engineering criteria as 'robustness' and 'stability' rather than by the physicist's method of looking for laws of nature. Gerald Sussman has also suggested that one should seek analogues of 'chaos' in complicated programming systems. However, in this paper we shall only deal with the simple textbook examples of programs, which seem to 'terminate' in a 'nonchaotic' way. Indeed, I shall present a discrete analogue of the Lyapunov asymptotic stability theory that is so useful in the theory of feedback control systems.

In mechanics, kinematics precedes dynamics. Kinematic structures often appear geometrically as fiber spaces. We shall see that the discrete analog of these geometric objects also appear naturally in some computer programs. In differential geometry, fiber spaces often have a group-theoretic nature, referred to as a fiber bundle with structure group. The discrete fiber spaces which occur here seem to carry a structure analogous to that of a fiber bundle: The algebraic notions of category theory [7, 8, 11, 18, 19] replace the role that the theory of transformation groups plays in differential-geometric fiber bundle theory. Such relations between category theory and the geometry of fiber spaces were pioneered in the work of Charles Ehresmann [7].

My aim in this paper is to set the stage for interaction between computer science, control theory, and differential and algebraic geometry, rather than to investigate the deeper aspects of the mathematical structures to be introduced here. In addition to developing mathematical formalism and structure, this paper will be concerned with explaining how some of the simpler algorithms to be found in the computer science literature can be efficiently described in these terms. We first deal with two sorts of algorithms, the *greatest common divisor* and *sorting* algorithms. We will show a parallel with control-theoretic ideas, particularly the classic *linear regulator* treated with 'Lyapunov' techniques. We then show how such Lyapunov techniques can be used more generally to find estimates of the 'complexity' of the computation, i.e. the number of steps required to complete it. We describe the geometric structure of algorithms which find the solution to first-order difference equations. In particular, the distinction between solution by 'recursion' and 'iteration' will be given an algebraic interpretation. Next, we define – by analogy with the geometric theory of differential systems – the notion of *prolongations* and *correspondences* between computer programs. We also go on to treat the computation of the binomial coefficients – and a generalization to what I call *triangularly recursive systems* – in terms of this formalism. The binomial coefficient algorithm is used as a main example in the work of Manna and Pneuli [21] on the use of temporal logic in the theory of program verification. Finally, we sketch a geometric way of thinking about modal logic which seems to be related to the way Manna and Pneuli use it for the description of certain aspects of programs.

The theory of *recursive functions* [5, 8] is, of course, one of the main foundational aspects of computer science, particularly those parts most closely related to the LISP language and its dialects. In conversations with Franklyn Turbak, the following way of thinking 'geometrically' about the way recursive functions are generated has suggested itself:

Let X and Y be sets, with

$$\phi\colon X \to Y \tag{1.1}$$

a map between them. In order to 'compute' ϕ, suppose given a set Z, two of its subsets Z' and Z'', and a pair (π', π'') of maps:

$$\pi'\colon Z' \to X, \tag{1.2}$$

$$\pi''\colon Z'' \to Y \tag{1.3}$$

satisfy the following condition:

$$\pi' \text{ is one-one and onto.} \tag{1.4}$$

Finally, suppose that for each $z' \in Z'$ one is given a subset:

$$T(z') \subset Z \tag{1.5}$$

called a *trajectory* through z'.

DEFINITION. The system $(Z, Z', Z'', \pi', \pi'', \{T(z')\colon z' \in Z'\})$ is said to *compute* the map ϕ if the following condition is satisfied:

$$\phi(\pi'(z')) = \pi''(T(z') \cap Z'') \quad \text{for all } z' \in Z'. \tag{1.6}$$

In the examples to be treated in this paper, the system $\{T(z')\colon z' \in Z'\}$ of trajectories is computed by solving a *recursion relation* or *difference equation*, thus linking up this 'geometric' structure with the classical theory of recursive functions [5, 8]. A special type of such a recursion relation might be generated by the action of the semi-group on Z, with the trajectories the orbits of the semi-group. In this paper, I will not pursue such a general framework, but will keep closer to some of the standard examples of computer programs to be found in books on computer programming.

2. Trajectories of Programs as Elements of Sheaves of Cross-Sections of Fiber Spaces

The textbook examples [1, 2, 3, 16, 23, 25] of computer programs – some of which will be considered below – seem to lead to the following sort of mathematical structure. First, there is a space T with a partial order relation $t < t'$ defined for pairs of elements of T. Think of 't' as some sort of 'time' coordinate of the program as it evolves. For each element t of T, we assume given a set $X(t)$, thought of as the collection of data at 'time' t describing the program at that 'instant'. (Computer scientists often use the term 'state' for what I have in mind, but I prefer to use that term in the more precise fashion it is used in control theory, as the 'Cauchy data', or information precisely required to determine the future time evolution of the system.)

The computer program – via the language-syntax-semantics process – will now produce a collection Γ of mappings with domains subsets of T and taking values in the spaces $X(t)$:

$$\gamma: t \to x(t) \in X(t), \tag{2.1}$$

that we will call the *trajectories* of the program. In order to give a geometric structure to this data, let us introduce the following construction:

Given the data $\{T, X(t) : t \in T\}$, construct another space, Z, in the following way:

$$Z = \{(x, t) : x \in X(t), t \in T\}. \tag{2.2}$$

Remark. This construction implicitly occurs in the computer science literature on *program verification and termination*. In this work [3, 9, 10, 13, 19, 20, 21], one assigns logical statements to 'pieces' of computer programs, and attempts to relate valuations of these statements at different points of the program. In geometric terms, one is looking at partial maps from subsets of the space Z constructed in (2.2) to other spaces, e.g. Boolean algebras, and then followed by valuation maps to other Boolean algebras. Mathematically, one deals with *sheaves* [11, 14] whose fibers are Boolean or Heyting algebras.

Define a mapping

$$\pi: Z \to T \tag{2.3}$$

by the formula

$$\pi((x, t)) = t, \quad \text{for } t \in T, \; x \in X(t). \tag{2.4}$$

π is a mapping of Z onto T. In geometric language, we will think of Z as a *fiber space* above the *base space* T, with the *projection map* π.

We then have

$$\pi^{-1}(t) = X(t), \quad \text{for } t \in T. \tag{2.5}$$

In the language of the geometry of fiber spaces, $X(t)$ is the *fiber of the fiber space* $\{Z, T, \pi\}$ above the point t of the base T.

A trajectory (2.1) can also be given a geometric name. Associate with a trajectory $t \to x(t) \in X(t)$ a mapping

$$\gamma: T \to Z \tag{2.6}$$

by the formula

$$\gamma(t) = (t, x(t)). \tag{2.7}$$

The following relation holds

$$\pi\gamma(t) = t, \quad \text{for all } t \in T. \tag{2.8}$$

In general, a map $\gamma: T \to Z$ with the property (2.8) is called a *cross-section* of the fiber space $\{Z, X, \pi\}$.

 ROBERT HERMANN

This construction motivates a sequence of definitions, in which we review certain standard geometric concepts

Let Z and T be sets, and let

$$\pi: Z \to T \tag{2.9}$$

be a map of Z onto T. The triple

$$(\pi, Z, T) \tag{2.10}$$

will be called a *fiber space*. For $t \in T$, the set

$$\pi^{-1}(t) = \{z \in Z, \pi(z) = t\} \tag{2.11}$$

will be called the *fiber of the fiber space* (2.10) above the point t.

DEFINITION. If $T' \subset T$, a map

$$\gamma: T' \to Z \tag{2.12}$$

is called a *cross-section map of the fiber space* (2.10) if the following condition is satisfied:

$$\pi\gamma(t) = t, \quad \text{for } t \in T'. \tag{2.13}$$

In other words,

$$\pi\gamma \text{ is the identity map } T' \to T'. \tag{2.14}$$

To see the geometric interpretation, for $t \in T$ set

$$X(t) = \pi^{-1}(t). \tag{2.15}$$

Condition (2.14) is equivalent to the following condition:

$$\text{For } t \in T, \gamma(t) \in X(t). \tag{2.16}$$

Thus, the map $t \to \gamma(t) \in X(t)$ picks out a point of the fiber above each point of the base space T.

Let us mention the sheaf-theoretic aspects of the assignment of a space of cross-section maps to the pair (T, π). Let $\mathbf{B}(T)$ be the Boolean algebra of subsets of T. $\mathbf{B}(T)$ has the usual Boolean operations – *union, intersection,* and *complementation* – and a *partial ordering* inclusion among the subsets. For $T' \in \mathbf{B}(T)$, set

$$\Gamma(T', \pi) = \text{set of cross-section maps } T' \to Z. \tag{2.17}$$

With π and T fixed, the assignment

$$T' \to \Gamma(T', \pi) \tag{2.18}$$

defines a map:

$$(\text{Subsets of } T) \to (\text{Partial maps } T \to Z) \tag{2.19}$$

which may be considered a pre-sheaf over T. [11, 14, 19].

Introduce a partial ordering on mapping spaces. It is extensively used in the Scott–Strachey approach to program semantics theory [19, 22, 24].

DEFINITION. Given two maps

$$\alpha\colon T' \to Z, \tag{2.20}$$

$$\beta\colon T'' \to Z, \tag{2.21}$$

with T', $T'' \subset T$, we say that $\alpha \leqslant \beta$ if and only if the following conditions are satisfied:

$$T' \subset T'', \tag{2.22}$$

$$\alpha = \beta \text{ restricted to } T'. \tag{2.23}$$

THEOREM 2.1. *The assignment* (2.19) *is a morphism of partially ordered sets, with* $\mathbf{B}(T)$ *given the Boolean partial ordering and the right-hand side of* (2.19) *given the partial ordering defined above.*

Proof. Straightforward verification. □

Let us now denote by

$$\Gamma'(\pi) = \{\gamma\colon T' \to Z\} \tag{2.24}$$

the space of partially-defined cross-sections, made into a partially ordered set as indicated above. As we shall see, the collection of trajectories of a computer program will define subsets Γ of $\Gamma'(\pi)$ with the following property:

$$\text{If } \gamma, \gamma' \in \Gamma'(\pi), \qquad \gamma \leqslant \gamma' \quad \text{and} \quad \gamma' \in \Gamma, \quad \text{then } \gamma \in \Gamma. \tag{2.25}$$

It will be convenient to give the notation

$$\Gamma(\pi) \tag{2.26}$$

to the space $\Gamma(T, \pi)$ of cross-sections of the fiber space which are globally defined over the base space T. $\Gamma'(\pi)$ can be reconstructed from $\mathbf{B}(T) \times \Gamma(\pi)$ as the quotient under the following equivalence relation:

$T' \times \gamma$ is equivalent to $T'' \times \gamma''$ if the following conditions are satisfied:

(a) $T' = T''$,
(b) $\gamma' = \gamma''$ on T'.

This representation of the space of partial cross-sections is convenient for various geometric purposes. For example, in differential geometry one deals with fiber spaces whose base and total space have manifold structures with the projection map π a smooth map. The pre-sheaf of locally-defined, smooth cross-sections can then be defined as the quotient of globally-defined smooth cross-sections, in the above way. This is especially convenient in the foundations of manifold theory, with Z just the product of T and the real numbers R. $\Gamma(\pi)$ is then the space of smooth, real-valued functions on the manifold T, from which all other associated differential-geometric structures of T can be reconstructed.

At this level of generality, we shall not attempt to describe the general properties of trajectories of computer programs beyond (2.25). Intuitively, they should have some sort of *recursive* structure. Perhaps – as in Scott's work [22, 24] – they should appear as fixed points of maps of $\Gamma'(\pi)$ into itself that have some continuity properties relative to topologies for $\Gamma'(\pi)$ that are compatible with the partial ordering. An alternate possibility is to work with some analog of the theory of recursive functions [5, 8].

Instead of pursuing this level of generality, we shall mainly restrict our attention to the case where the 'dynamics' is prescribed by difference equations. Indeed, our main interest will be in the algebraic structure of these difference equations. We will assume that the base space T has a distinguished operator δ that will play the part of a time-translation operator. The main example in practice is the case where T is the set of nonnegative integers with the usual ordering, but it is worthwhile – perhaps for ultimate application to the theory of parallel and concurrent processes – to work in the more general situation. We then exhibit some textbook examples of computer programs as prototypes for the introduction of mathematical structures akin to those encountered in control theory and geometry.

3. The Kinematic and Recursive Structure of Programs of the von Neumann Type

Much of the traditional von Neumann style of computer programming is based on the mathematics of recursive functions, as developed in the 1930s work of Gödel, Church, Turing, and Kleene. This classic work has served as a model for computer science, and was originally developed in the context of functions on the nonnegative integers, a set denoted as $\mathbf{N}$, which we will use as a base space for our 'kinematic' fiber space

$$\pi: Z \to \mathbf{N}. \tag{3.1}$$

An element of $\Gamma(\pi)$ is then more familiarly thought of as a sequence

$$n \to x(n) \in X(n) = \pi^{-1}(n) \tag{3.2}$$

of elements of the fibers of the fiber space (3.1). (The reader should keep in mind that the sequence (3.2) might only be defined for a subset of the integers.) A subset $\Gamma \subset \Gamma(\pi)$ might be the trajectories of a computer program if it consists of *recursive* objects. See the book by Eilenberg and Elgot [8] for indications of how one might define this concept in greater generality. Instead of working here at such a level of generality, we shall deal in this paper with the following more elementary concept of 'recursiveness':

For each integer n, and for fixed integer k and for each

$$(x(n), \ldots, x(n - k)) \in X(n) \times \cdots \times X(n - k),$$

suppose, given a map,

$$f_n(x(n), \ldots, x(n - k)): X(n) \times \cdots \times X(n - k) \to X(n + 1). \tag{3.3}$$

Let us suppose then that the trajectories

$$\gamma = (x(0), x(1), \ldots) \in \Gamma$$

satisfy the following difference equation:

$$x(n + 1) = f_n(x(n), \ldots, x(n - k))((x(n), \ldots, x(n - k)) \tag{3.4}$$

The difference equations (3.4) then determine the 'evolution' of the program trajectories.

Of course, in practice the main problem is the choice of the system of maps f_n, to achieve some set of pre-assigned conditions, and perhaps also achieve termination at some finite integer that will not grow too large as the given data increases in size. Control theory is the branch of applied mathematics and engineering that most closely concerns itself with problems of this type. Indeed, we will see that some of the standard algorithms of computer programming theory have an algebraic structure closely paralleling that of feedback control theory. Another possibility to keep in mind is that the f_n may themselves be defined by recursion relations: This approach would be in the spirit of the classical theory of recursive functions.

Next, we turn to the study of two of the GCD algorithms of computer science textbooks in order to bring the foreground a special algebraic structure of recursion relations/differential equations of the form (3.4).

4. The Fiber Bundle Structure of the Greatest Common Divisor Algorithms

In this section, we shall consider the two GCD algorithms commonly used as examples of recursive computations in computer science textbooks [1, 2, 3, 23]. The fiber space $\pi: Z \to N$ associated with these algorithms will be constructed in the following way: The base space is N, i.e. the programs are of the von Neumann type. Let Z denote the set of all integers, with N embedded as the set of nonnegative integers. Set

$$Z = \{(n, (j, k)\}, \tag{4.1}$$

with $n \in N, j$ and $k \in Z$. Let the fiber space projection map be defined as

$$\pi(n, (j, k)) = n, \quad \text{for } n \in N, j \quad \text{and} \quad k \in Z. \tag{4.2}$$

Let $L: Z \to Z$ be the map defined by the following formula:

$$L(n, (j, k)) = \text{minimum}(j, k) \tag{4.3}$$

The cross-section maps $\Gamma(\pi) = \{\gamma\}$ are then of the following form:

$$\gamma(n) = (n, (j(n), k(n))), \quad \text{for } n \in N. \tag{4.4}$$

Either one of the standard GCD algorithms defines – starting with a given pair (j, k) of nonnegative integers – a sequence

$$(j(1), k(1)), (j(2), k(2)), \ldots$$

of pairs of nonnegative integers whose GCD is independent of n and which ends after a finite number of steps with a pair of integers one of whose components is *zero*.

Remark. This is a special case of the idea that certain computational algorithms involve finding – by a finite, Noetherian, process – a *canonical form* for the expressions to be computed.

Since the GCD of $(m, 0)$ or $(0, m)$ is readily evaluated, and $(j(n + 1), k(n + 1))$ is given in terms of $(j(n), k(n))$ by an explicit, readily-evaluated formula, putting these facts together provides the computation of the GCD of (j, k).

Let us now make the following definition, which proposes a general, 'geometric', description of desirable properties for a GCD algorithm:

DEFINITION. A *trajectory system of a GCD algorithm* is a subset $\Gamma = \{\gamma: n \rightarrow (j(n), k(n))\}$ of the set $\Gamma(\pi)$ with the following properties:

(a) $j(n)$ and $k(n) \in \mathbf{N}$,   for all $n \in \mathbf{N}$, all $\gamma \in \Gamma$. $\qquad\qquad\qquad\qquad$ (4.5)

(b) The initial value map $\gamma \rightarrow (j(0), k(0))$ is a
     ~~one-~~one map of Γ onto $\mathbf{N} \times \mathbf{N}$. $\qquad\qquad\qquad\qquad\qquad\qquad$ (4.6)

(c) For each $\gamma \in \Gamma$, the map $n \rightarrow \mathrm{GCD}(j(n), k(n))$ is constant. $\qquad\qquad$ (4.7)

(d) For each $\gamma \in \Gamma$, the map $n \rightarrow L(\gamma(n))$ is a decreasing function of n. $\qquad$ (4.8)

The following theorem shows how such a trajectory system can be used to compute the greatest common divisor of a given pair (j, k) of integers:

THEOREM 4.1. *Let Γ be a given* GCD *trajectory system, satisfying conditions* (4.5)– (4.8). *Let* (j, k) *be a given pair of nonnegative integers. Let γ be an element of Γ that satisfies the following conditions*:

$$\gamma(0) = (j, k), \qquad\qquad\qquad\qquad\qquad\qquad\qquad\qquad\qquad (4.9)$$

$$n \rightarrow L(\gamma(n)) \text{ is a strictly decreasing function.} \qquad\qquad\qquad (4.10)$$

Then, there is an $n \in \mathbf{N}$ such that

$$\gamma(n) = (\mathrm{GCD}(j, k), 0) \quad or \quad (0, \mathrm{GCD}(j, k)). \qquad\qquad\qquad (4.11)$$

Thus, if $\gamma(0)$, $\gamma(1), \ldots$ are the sequences of results of a von Neumann type of computer program, then $\gamma(n)$ 'computes' the GCD *of the pair* (j, k).

Proof. Follows immediately from (4.5)–(4.10). $\qquad\qquad\qquad\qquad\qquad\qquad$ $\square$

Remark. This argument also contains within it a version of a very useful and powerful control theoretic-idea for assuring correct termination, which in essence goes back to Lyapunov's classical work of stability of differential equations. In turn, this method is based on the idea from physics of a conservative or dissipative system, namely a physical system (described in the usual way by trajectories of differential equations) whose trajectories have a real-valued function – perhaps to be identified with energy – which remains constant (in the conservative case) or strictly decreases with time (in the dissipative situation). In control theory, the Lyapunov method has been most strikingly used by Kalman in his now-classical treatment of feedback stabilization of the linear quadratic regulator (most accessibly described in the book by Anderson and Moore [4].

Notice that the GCD algorithm described above is a variant of the traditional Lyapunov argument. We deal with discrete-time dynamical systems on the space of pairs (j, k) of integers such that the following real-valued function:

$$(j, k) \to \text{minimum}(j, k) \tag{4.12}$$

strictly decreases with increasing 'time'. Also, the following condition is satisfied:

$$\text{The real-valued function } (j, k) \to \text{GCD}(j, k) \text{ is } constant \text{ in 'time'.} \tag{4.13}$$

We deduce that in finite 'time', i.e. in a finite number of steps of the algorithm, the process terminates with a pair (j, k) satisfying the following conditions:

$$n = 0, \qquad \text{GCD}(j, k) = j, \tag{4.14}$$

thus completing successfully the algorithm.

In the computer science literature, there are two methods commonly used to generate a family of trajectories satisfying (4.5)–(4.8). Both define $\gamma(n + 1)$ recursively in terms of $\gamma(n)$, in terms of a feedback control structure. We shall deal first with the one that relies on the division algorithm for integers.

Let us exploit the algebraic structure of the fiber space π. Note that each fiber is identified with $\mathbf{Z} \times \mathbf{Z}$, hence has the structure of an Abelian group, denoted additively. The set of homomorphisms of this group into itself is a *monoid*, i.e. has a binary product structure which is associative and has a unit element. This monoid – which we denote as G – can be made explicit as the set of 2×2 matrices with integer coefficients, in the following way:

Given a quadruple (a, b, c, d) of elements of $\mathbf{Z}$, we define a map

$$g(a, b, c, d) \colon \mathbf{Z} \times \mathbf{Z} \to \mathbf{Z} \to \mathbf{Z}$$

by the following formula:

$$g(a, b, c, d)(j, k) = (aj + bk, cj + dk). \tag{4.15}$$

THEOREM 4.4. *Given $(j, k) \in \mathbf{N} \times \mathbf{N}$, with $k < j$, there is a $g \in G$ (constructed using the Euclidean division algorithm) such that the image*

$$(j', k') = g(j, k) \tag{4.16}$$

of (j, k) under g satisfies the following conditions:

$$\text{GCD}(j, k) = \text{GCD}(j', k'), \tag{4.17}$$

$$k' < k, \tag{4.18}$$

$$L(j', k') < L(j, k). \tag{4.19}$$

Proof. Using the division algorithm, we can write

$$j = q(j, k)k + r(j, k), \tag{4.20}$$

with the following conditions satisfied

$$q(j, k), r(j, k) \in \mathbf{N}, \tag{4.21}$$

$$r(j, k) < k. \tag{4.22}$$

The integers $q(j, k)$, $r(j, k)$ are uniquely determined by conditions (4.20)–(4.22), i.e. the assignments $(j, k) \to q(j, k)$, $r(j, k)$ define maps of $\mathbf{N} \times \mathbf{N} \to \mathbf{N}$. Set

$$j' = k, \tag{4.23}$$

$$k' = r(j, k). \tag{4.24}$$

We now show that (j', k') has the same GCD as (j, k). Let d divide j and k. We see from (4.20), (4.23) and (4.24) that d divides j' and k'. Suppose, conversely, that d' divides j' and k'. From (4.17) we see that d' divides n. (4.24) implies that d' divides q. (4.20) implies that d' divides m. This implies (4.17). $\qquad \square$

This algebraic result is the foundation of one of the standard computer science GCD algorithm. To explain how this algorithm may be described in more algebraic terms (and in Control Theory terms as a closed-loop trajectory of a feedback system), proceed as follows:

For (j, k), $(l, m) \in \mathbf{N} \times \mathbf{N}$, let $g(j, k)$ be the element of G defined by the following formula:

$$g(j, k)(l, m) = (m, l - q(j, k)m) = (m, r(j, k)) \tag{4.25}$$

where $q(j, k)$ and $r(j, k)$ are defined by (4.20).

In matrix terms,

$$g(j, k) = \begin{pmatrix} 0 & 1 \\ 1 & q(j, k) \end{pmatrix}. \tag{4.26}$$

Let us now use this algebraic structure to construct the trajectories of one of the standard GCD algorithms:

THEOREM 4.3. *Let $\pi\colon Z \to \mathbf{N}$ be the fiber space constructed in (4.1). Let $\mathbf{f}\colon Z \to Z$ be the map defined by the following formula:*

$$\mathbf{f}(n, (j, k)) = (n + 1, f_n(j, k)), \tag{4.27}$$

with

$$f_n(j, k) = (k, r(j, k)) = g(j, k)(j, k), \tag{4.28}$$

where $(j, k) \to g(j, k)$ is the map of $\mathbf{N} \times \mathbf{N}$ to G defined by formula (4.25).

Given $(0, (j, k)) \in Z$, with $0 < k < j$, let

$$n \to (j(n), k(n))$$

be the solution of the following difference equation:

$$(j(n + 1, k(n + 1)) = f_n(j(k), k(n)), \tag{4.29}$$

with the following initial condition:

$$j(0) = j, \qquad k(0) = k.\tag{4.30}$$

Then, for n sufficiently large,

$$j(n) = \mathrm{GCD}(j, k), \qquad k(n) = 0.\tag{4.31}$$

Thus, the corresponding cross-section map $n \to g(n) = (n,(j(n), k(n)))$ provides a trajectory for the first GCD algorithm.

The algorithm just presented for the GCD requires having available the algorithm for the division of one integer by another. The other GCD algorithm in the computer science literature does not use prior knowledge of a division algorithm. Instead, it uses successive subtractions. Its algebraic simplicity is balanced by a greater combinatoric complexity, since it involves a feedback which depends on a logical decision, which may switch in a complicated way. This is, in fact, a simple prototype of a phenomenon which is often encountered in practice, but involving a much more complicated and elaborate feedback switching structure. Let us present a few brief remarks about this version of the GCD algorithm.

THEOREM 4.4. *If $(j, k), (j', k') \in \mathbf{N} \times \mathbf{N}$ and the following relations are satisfied:*

$$(j', k') = (j, k - j) \quad or \quad (j - k, k),\tag{4.32}$$

then

$$\mathrm{GCD}(j, k) = \mathrm{GCD}(j', k')\tag{4.33}$$

Proof. If d divides j and k then it obviously divides $j, k, k - j$, and $j - k$. Conversely, anything dividing j and $k - j$ divides k also. Formula (4.33) follows. $\qquad\square$

This result gives us a method for computing the $\mathrm{GCD}(j, k)$, and a strategy for a method of construction of a closed-loop feedback control law whose trajectories at a given starting point (j, k) in $\mathbf{N} \times \mathbf{N}$, will end up at a finite 'time' at a point of the following subset:

$$\{(j, k) \in \mathbf{N} \times \mathbf{N} : j = 0 \text{ or } k = 0\}.$$

Namely, if one is at the point (j, k) of $\mathbf{N} \times \mathbf{N}$, proceed at the next step to $(j, j - k)$ or $(k, k - j)$ depending on a 'logical' decision to be made, so as to stay in $\mathbf{N} \times \mathbf{N}$.

Let us formulate this procedure more systematically. Set $\mathbf{B} = \{\mathbf{T}, \mathbf{F}\}$, the two-element Boolean algebra, i.e. the algebra of subsets of a space with one element, with '$\mathbf{T}$' identified with the whole space and with '$\mathbf{F}$' identified with the empty set. Let $\beta' : \mathbf{N} \times \mathbf{N} \to \mathbf{B}$ be the map defined in the following way:

$$\beta'(j, k) = \mathbf{T}, \quad \text{if } `j - k \geqslant 0\text{' is true},\tag{4.34}$$

$$\beta'(j, k) = \mathbf{F}, \quad \text{if } `j - k \geqslant 0\text{' is false}.\tag{4.35}$$

Let $\beta'' : \mathbf{B} \to G$ be the map defined by the following formula:

$$\beta''(\mathbf{T}) = g(0, 1, 1, -1) = g_+,\tag{4.36}$$

$$\beta''(\mathbf{F}) = g(1, 0, -1, 1) = g_-.\tag{4.37}$$

Set

$$\beta = \beta''\beta'. \tag{4.38}$$

It is a map: $\mathbf{N} \times \mathbf{N} \to G$.

THEOREM 4.5. *Let* $(j, k) \in \mathbf{N} \times \mathbf{N}$, *and let* $n \to (j(n), k(n))$ *be the solution of the following difference equations*

$$(j(n + 1), k(n + 1) = \beta(j(n), k(n))(j(n), k(n)). \tag{4.39}$$

Then, $n \to (n, (j(n), k(n)) = \gamma(n)$ *is a trajectory on which* L *is strictly decreasing and on which* GCD *is constant, hence terminates to provide an algorithm for the computation of* $\mathrm{GCD}(j, k)$.

Proof. Let us first suppose that

$$j(n) \geqslant k(n). \tag{4.40}$$

Using (4.34),

$$\begin{aligned}
\beta(j(n), k(n))(j(n), k(n)) &= \beta''\beta'((j(n)), k(n))(j(n), k(n)) \\
&= g_+(j(n), k(n)) \\
&= (k(n), j(n) - k(n)).
\end{aligned}$$

The complementary case to (4.40) is handled similarly. $\qquad\square$

5. Categories as the Abstract Theory of Fiber Preserving Mappings ('Gauge Transformations') of Fiber Spaces

The algebraic structures we call *groups* play a fundamental role in both pure and applied mathematics. Historically, they arose as abstractions of the sort of algebraic structure encountered in studying collections of mappings of a set into itself that are closed under composition and inverses. Categories are algebraic structures which extend the 'group' notion: They arose in the 1940's as abstraction of material from algebraic topology. However, they also appear in a fundamental way in differential geometry as algebraic structures acting on fiber spaces as 'fiber-preserving maps' or – as the physicists like to call them – 'gauge transformations'. This geometric interpretation of category theory was first emphasized in the seminal work of Charles Ehresmann [7] during the 1950's and 1960's. The now-standard algebraic foundational approach to the category concepts may conveniently be found in [8, 11, 18].

My ultimate purpose here is to show how categories acting on fiber spaces can be used to prescribe the 'dynamics' of computer programs. In this section, I will first show how the fiber space intuition can be used to motivate the abstract definition of a category.

Let

$$\pi: Z \to T \tag{5.1}$$

be a fiber space, as before. For $t \in T$, let

$$X(t) = \pi^{-1}(t) \tag{5.2}$$

be the fiber above the point t.

For $(t, t') \in T \times T$, let:

$$M(t, t') = \text{set of maps of } X(t') \to X(t). \tag{5.3}$$

Consider the set

$$\mathbf{M} = \{M(t, t') : (t, t') \in T \times T\} \tag{5.4}$$

Each element of $\mathbf{M}$ is a *partially-defined* mapping from Z to Z. Hence, composition of mappings defines a partially-defined binary algebraic structure on $\mathbf{M}$. This binary product is *associative*, wherever it is defined. It has not just one *unit* element, but a whole family: The identity maps of each of the fibers $X(t)$ into itself.

Abstracting from this example leads to the following definition.

DEFINITION. A *category* is a composite algebraic and geometric structure – denoted by a single bold-face letter such as $\mathbf{G}$ – defined by giving the following data:

(a) A set T called the *objects* of the category.

(b) A map $\alpha : \mathbf{G} \to T \times T$. The elements of $\mathbf{G}$ are called the *morphisms* of the category.

For $(t, t') \in T \times T$, let

$$\mathbf{G}(t, t') = \alpha^{-1}(t, t') \tag{5.5}$$

be the fiber of the map α above the point (t, t'). For each triple (t, t', t''), we postulate a binary mapping

$$\mathbf{G}(t, t') \times \mathbf{G}(t', t'') \to \mathbf{G}(t, t''), \qquad (g, g') \to gg', \tag{5.6}$$

satisfying the following version of the *associative law*:

For each quadruple $(t, t', t'', t''') \in T \times T \times T \times T$, each triple

$$(g, g', g'') \in \mathbf{G}(t, t') \times \mathbf{G}(t', t'') \times \mathbf{G}(t'', t'''), \tag{5.7}$$

we have

$$(gg')g'' = g(g'g'') \tag{5.8}$$

Finally, to complete the prescription for the definition of a category, we postulate the existence of distinguished elements with properties generalizing the 'unit element' of a group:

For each $t \in T$, we are given an element

$$\mathbf{l}(t) \in \mathbf{G}(t, t) \tag{5.9}$$

such that the following conditions are satisfied:

$$\mathbf{l}(t)g = g, \quad \text{for all } g \in \mathbf{G}(t, t'), \tag{5.10}$$

$$g\mathbf{l}(t') = g, \quad \text{for all } g \in \mathbf{G}(t, t'). \tag{5.11}$$

Having defined a category in this way as an algebraic structure like any other (but with an additional 'geometric' structure), we could now define by analogy such familiar algebraic notions as 'homomorphism' and 'isomorphism', that are traditionally associated with the group concept. However, the standard terminology in the category literature for these concepts is not in accordance with this intuition, and we shall not pursue it here. (For example, the analogue of a 'homomorphism' is called a *functor*.) To close this brief excursion into general algebra, let us define the analogue of *transformation group*. This concept is the most important one for the applications we have in mind.

DEFINITION. Let **G** be a category and let $\pi: Z \to T$, be a fiber space, with $X(t) = \pi^{-1}(t)$ for $t \in T$. **G** is said to *act* on the fiber space if the following conditions are satisfied:

T is the set of objects of **G**. $\hfill (5.12)$

For each $(t, t') \in T \times T$, there is a map

$$\mathbf{G}(t, t') \times X(t') \to X(t), \qquad (g, x) \to gx, \hfill (5.13)$$

for $g \in \mathbf{G}(t, t')$, $x \in X(t')$.

The following compatibility condition is satisfied between the partially-defined binary product on **G** and the product (5.13)

$$g'(gx) = (g'g)x, \quad \text{for } g \in \mathbf{G}(t, t'), \, g' \in \mathbf{G}(t'', t), \, x \in X(t'). \hfill (5.14)$$

Let us call this the *associative law*.

Let us now make explicit how such an action defines a family of fiber-preserving partial maps on the fiber space Z.

THEOREM 5.1. *Suppose that the category* **G** *acts on the fiber space* π, *as explained above. Then for each* $g \in \mathbf{G}(t, t')$, *let* $\phi_g: X(t') \to X(t)$ *be the map defined by the formula*

$$\phi_g(x) = gx, \quad \text{for } x \in X(t'). \hfill (5.15)$$

The associative law translates into the following composition law for the mappings $\{\phi_g: g \in \mathbf{A}\}$:
For $g \in \mathbf{G}(t, t')$ *and* $g' \in \mathbf{G}(t'', t)$

$$\phi_{gg'}(x) = \phi_g(\phi_{g'}(x)), \quad \text{for } x \in X(t'). \hfill (5.16)$$

Proof. That the assignment $g \to \phi_g$ maps $X(t')$ into $X(t)$ follows from (5.13). To prove (5.16), note that left-hand side is $(gg')(x)$, while the right-hand side is $g(g'x)$. Equality follows from (5.12). $\hfill \square$

6. Recursive Feedback Dynamics Associated with Categories

Group theory is associated with the study of symmetries associated with differential equations. In a similar spirit, I now want to present methods for the description of the recursive dynamics of computer algorithm trajectories in terms of categories acting on fiber spaces, as described in the previous section.

Let

$$\pi: Z \to T \tag{6.1}$$

be a fiber space, with

$$t \to X(t) = \pi^{-1}(t) \tag{6.2}$$

the following mapping:

$$\text{Base space } T \to (\text{Subspaces of } Z \text{ that are fibers of } \pi). \tag{6.3}$$

Recall that we have defined $\Gamma(\pi)$ as the space of all cross-sections of the fiber space π.

Let $\mathbf{G}$ be a category with objects T. Let us suppose that this category acts on the fiber space T, i.e. a mapping:

$$\mathbf{G}(t, t') \times X(t') \to X(t)$$

$$(g, x) \to gx \tag{6.4}$$

is defined which is associative relative to the product within the category.

In order to define a structure on T which will allow iteration to take place, let us suppose that we are given a map

$$\delta: T \to T. \tag{6.5}$$

It will be called an *iteration operator*.

Remark. The case to keep in mind is

$$T = \mathbf{N}, \tag{6.6}$$

$$\delta(n) = n + 1, \quad \text{for } n \in \mathbf{N} \tag{6.7}$$

This choice leads us to the classical theory of difference operators, equations, and solution of these by iteration. Mathematicians tend to use the terms 'iteration' and 'recursion' interchangeably, whereas a precise distinction between them is essential for computer science purposes. I will explain one way of differentiating between them in the next section.

Among the possible generalizations, perhaps the most important is the case where T is a semigroup (denoted additively) and δ is the *translation* with respect to a fixed element t_0 of the semigroup:

$$\delta(t) = t + t_0. \tag{6.8}$$

Applying the recursion theory sketched in this and the following sections, also covers certain systems of *partial difference equations*.

As we have seen for the example of the GCD algorithms, there is often a feedback structure associated with the common computer algorithms. Here is one version of this ubiquitous idea:

DEFINITION. A *feedback law* (relative to the operator δ and category $\mathbf{G}$) is a mapping:

$$K: Z \to \mathbf{G} \tag{6.9}$$

satisfying the following condition

$$K(t, x) \in \mathbf{G}(t, \delta t), \quad \text{for } t \in T, \ x \in X(t). \tag{6.10}$$

We now want to use such a mapping K to define an equation for cross-sections of the fiber space π. Let

$$\gamma \colon T \to Z \tag{6.11}$$

be a cross-section map. It has the form

$$\gamma(t) = (t, x(t)). \quad \text{for all } t \in T, \text{ with } x(t) \in X(t). \tag{6.12}$$

DEFINITION. γ is said to be a *trajectory* of the feedback law K if the following equation is satisfied:

$$K(t, x(t))(t, x(t)) = (\delta(t), x(\delta t)) \tag{6.13}$$

for all $t \in T$.

Notice that (6.13) is a difference equation for the map $t \to x(t)$. We have seen how the trajectories of the GCD algorithm are constructed in this way from a feedback law associated with the category of linear fiber-preserving maps of the fiber space

$$\mathbf{N} \times (\mathbf{Z} \times \mathbf{Z})$$

into itself. Below, we shall see how *sorting algorithms* are constructed from categories associated with the permutation groups.

Let us now study general difference equations constructed using the operator δ.

7. Difference Equations, Fixed Points, Iteration and Recursion

Many of the algorithms found in computer science books are constructed as simple iterative or recursive solutions of difference equations, with no feedback involved at all. The traditional example of this type is the factorial function of the nonnegative integers into itself. We shall now discuss a general formulation of this simplest programming situation.

Let $\pi \colon Z \to T$ be a fiber space map, with $X(t) = \pi^{-1}(t)$ the fiber above the point $t \in T$. Let $\delta \colon T \to T$ continue as in the previous section. Suppose that

$$f \colon Z \to Z \tag{7.1}$$

is a map such that

$$f(X(t)) \subset X(\delta t), \quad \text{for all } t \in T. \tag{7.2}$$

Remark. Geometrically, (7.2) says that f is a fiber-preserving mapping. δ then is the map on the base space which keeps track of which fibers are mapped into which by f.

We can associate with the pair (δ, f) the following *difference equation*:

$$\gamma(\delta(t)) = f(\gamma(t)). \tag{7.3}$$

A *solution* is a cross-section map $\gamma \colon T \to Z$ that is defined and satisfies (7.3) for all $t \in T$.

Let us now describe one straightforward method for computing the value $\gamma(\delta^n(t))$ of the solution γ of the difference equation (7.3).

THEOREM 7.1. *Let γ be a solution of the difference equation (7.3). Then, for each $t \in T$, the following formula holds:*

$$\gamma(\delta^n(t)) = f(f(\cdots(\gamma(t)\cdots))). \tag{7.4}$$

This formula may be called the *iterative formula for the value at $\delta^n(t)$ of the solution γ of the difference equation* (7.3). Algebraically, it expresses the value of a solution as an *orbit* of the action of the category (or 'partially defined' semi-group) generated by the map f.

Proof. Use (7.3) twice:

$$\gamma(\delta^2(t)) = \gamma(\delta(\delta(t))) = f(\gamma\delta(t)) = f(f(\gamma(t))$$

Continue by induction on n to prove formula (7.4). □

Remark. This is also an example of what the computer scientists call a *rewrite rule*: The formula on the left-hand side of (7.4) is 'rewritten' using (7.3) as an 'axiom' until the right-hand side is obtained. Many automated deductions and algebraic computations involve this sort of reasoning.

In order to describe in mathematical terms, the reasoning that goes into what computer scientists call the *recursive solution* to the difference equation, we will now define a map α on the space of cross-sections into itself such that γ is a solution of the difference equation if and only if it is a fixed point of α. To do this, let us consider a map

$$\rho: \delta(T) \to T \tag{7.5}$$

which satisfies the following condition:

$$\rho\delta(t) = t, \quad \text{for all } t \in T. \tag{7.6}$$

DEFINITION. ρ will be called a *recursion map* associated to the *iteration map δ*.

For example, in the case where: $T = \mathbf{N}$, $\delta(n) = n + 1$, then $\rho(n) = n - 1$.

One might say then that *iteration goes forward, recursion goes backward*. Notice that ρ is not uniquely defined by condition (7.6). In fact, the set of such ρ's parametrizes the solutions of the difference equation.

Return to the general case and set

$$\alpha(\gamma)(t) = \begin{cases} f(\gamma(\rho(t))), & \text{for } t \in \delta(T), \\ \gamma(t), & \text{for } t \in T - \delta(T). \end{cases} \tag{7.7}$$

THEOREM 7.2. *Let $\Gamma(\pi)$ be the space of globally-defined cross-section of the fiber space π. Then, α defined by (7.7) is a map $\Gamma(\pi) \to \Gamma(\pi)$. Also,*

$$\alpha(\gamma) = \gamma \tag{7.8}$$

if and only if γ is a solution of the difference equation (7.3).

Proof. Suppose first that (7.8) is satisfied. Then, for $t \in T$,

$$\gamma(\delta(t)) = \alpha(\gamma)(\delta(t))$$
$$= f(g(\rho\delta(t))), \quad \text{using (7.7),}$$
$$= f(\gamma(t)), \quad \text{using (7.6).}$$

which prove that γ is a solution of the difference equation (7.3). The converse is proved by reversing the argument. $\qquad\square$

Let us now see how the fixed point functional equation (7.8) of the solution f of the difference equation can be iterated to provide a method for computing the solution to the difference equation (7.3). This process provides what the computer scientists call the *recursive solution* to the difference equation (7.3).

Iterating (7.8) gives

$$\alpha^n(\gamma) = \gamma. \tag{7.9}$$

Of course, this is just another form (or 'rewrite') of formula (7.4). Let us evaluate both sides of (7.9) at a point $t \in T$:

$$\gamma(t) = \alpha(\alpha^{n-1}(\gamma))(t) = f((\alpha^{n-1}(\gamma))(\rho(t))).$$
$$= \cdots. \tag{7.10}$$

For example, for $n = 2$

$$\gamma(t) = f(\alpha(\gamma)(\rho(t))) = f(f(\gamma)(\rho^2(t))). \tag{7.11}$$

Formulas like (7.10) or (7.11) may be said to have a 'recursive' form because they go 'backwards' in 'time' via the operator ρ rather than 'forwards' in 'time' via the iteration operator δ.

The description in terms of trajectories of the iterative and recursive versions of the solution-of-first-order-difference-equations will be left to the reader. Notice that the iterative version involves sequences of points of the fiber space π while the recursive version involves sequences of elements of the space of cross-sections of this fiber space.

8. Lyapunov Methods for Estimation of Complexity of Computer Programs

Let us now turn to the general formulation of the 'Lyapunov' ideas that we have encountered in our study of the special case of the GCD algorithm. Let $\pi: Z \to T$, $t \to X(t) = \pi^{-1}(t)$, be a fiber space, and let

$$\gamma: T \to Z, \quad t \to (t, x(t)) \tag{8.1}$$

be a cross-section of π.

Let

$$\delta: T \to T \tag{8.2}$$

be a map which will serve as an iteration operator to construct difference equations for cross-sections of π.

DEFINITION. A map

$$\mathbf{L}\colon Z \to \mathbf{N} \tag{8.3}$$

is said to satisfy the *Lyapunov condition* relative to the cross-section (8.1) and iteration operator (8.2) if the following condition is satisfied:

$$\mathbf{L}(\gamma(\delta t)) \leqslant \mathbf{L}(\gamma(t)), \quad \text{for all } r \tag{8.4}$$

It is said to satisfy this condition *strictly* if the symbol $\leqslant$ in (8.4) can be replaced by strict inequality '$<$'. $\mathbf{L}$ is said to satisfy the Lyapunov condition relative to a difference equation (or discrete-time flow) if it satisfies condition (8.4) for every solution of the difference equation, i.e. every trajectory of the flow.

Let us now make precise what the existence of such a Lyapunov map says about the length of 'time' spent before the cross-section (8.3) hits a given goal.

THEOREM 8.1. *Suppose that* $\mathbf{L}$ *satisfies the strict Lyapunov condition relative to the cross-section* γ, *that* Y *is a subset of* Z, t_0 *is point of* T, *and that the following terminal conditions are satisfied:*

$$\gamma(\delta^k(t_0)) \in Y, \quad \text{with } k \in \mathbf{N}. \tag{8.5}$$

$$\mathbf{L}(Y) = 0. \tag{8.6}$$

Then, the following upper bound for k *holds:*

$$\mathbf{L}(\gamma(0)) \geqslant \kappa \tag{8.7}$$

Proof. We have

$$0 = \mathbf{L}(\gamma(\delta^k(t_0))) \geqslant \mathbf{L}(\gamma(\delta^{k-1}(t_0))) + 1 \geqslant \ldots, \tag{8.8}$$

hence (8.7) follows from (8.5) and (8.6). $\qquad\square$

In computer science terms, if the cross-section γ arises as a result of a computation of a computer program in von Neumann form, notice that (8.7) gives us an estimate of the length of the program. Complexity theory, in the computer science sense, often deals with asymptotic estimates for this number as certain parameters in the initial value $\gamma(t_0)$ vary. For example, in the GCD algorithm, $\gamma(t_0)$ is the initial pair (j, k) of integers, and $\mathbf{L}(\gamma(t_0))$ is the minimum of j and k.

Let us now formalize how Lyapunov ideas may be used to construct closed-loop feedback laws. Let us now suppose that the base space T is the nonnegative integers $\mathbf{N}$. Suppose given a discrete-time, open-loop Control System:

$$x(j + 1) = f_j(x(j), u(j)), \quad x(j) \in X(j), \, u(j) \in U(j). \tag{8.9}$$

Suppose that we are given an initial point x_0 and a 'target' subset S of $Z = \{(j, x(j)\colon j \in \mathbf{N}, x(j) \in X(j)$. Our aim is to construct feedback maps:

$$K_j\colon X_j \to U_j \tag{8.10}$$

such that solution $j \to x(j)$ of the following closed-loop iteration equations:

$$x(j + 1) = f_j(x(j)), \, K_j(x(j)) \tag{8.11}$$

satisfies the conditions

$$x(0) = x_0, \tag{8.12}$$

$$x(k) \in S, \quad \text{for some integer } k. \tag{8.13}$$

Our strategy is to first choose a potential Lyapunov function

$$\mathbf{L}: Z \to \mathbf{N} \tag{8.14}$$

such that

$$\mathbf{L}(S) = 0, \tag{8.15}$$

and then to choose K so as to solve the inequality condition

$$\mathbf{L}(j, f_j(x(j)), K(x(j))) < \mathbf{L}(x(j)), \quad \text{for all } x(j) \in X(j). \tag{8.16}$$

Sufficient conditions for such extremal equations to have solutions are not obvious, just as they are not in the analogous dynamical systems case. For example, note that the existence of such a Lyapunov function guarantees termination of the algorithm.

What conditions on f_j would result in the existence of such a pair $(\mathbf{L}, K)$ of maps is also not obvious in this generality. Here is where one might expect geometric intuition to be useful. A starting point might be to choose $\mathbf{L}$ to be a function of the distance to the target subset S, in terms of a metric on the state space X.

9. Sorting Algorithms from the Feedback Control Point of View

We now turn to the consideration of the *sorting algorithms* to provide further examples where the algebraic structure introduced above seems to be naturally associated with important computer algorithms. We shall limit our attention here to the *bubble sort* and *quick sort* algorithms, since the algebraic nature of these algorithms is particularly striking.

Let us summarize standard material about the permutation group. Let k be a fixed integer, and let S be a finite set with k elements. Let G be the group of one-one, onto maps $g: S \to S$. G is isomorphic to the permutation group on k elements.

DEFINITION. A one-one map

$$\kappa: S \to R \tag{9.1}$$

is called a *key* for the set S.

Following the treatment in Knuth's third volume [15], let us formulate the sorting problem as follows:

DEFINITION. A map

$$\phi_k: \{1, \ldots, k\} \to S \tag{9.2}$$

is called a *sorting map* for the set S and key (9.1) if the following conditions are satisfied:

(a) ϕ_k is one-one (and therefore also onto). $\tag{9.3}$

(b) The composite map

$$k\phi_k: \{1, \ldots, k\} \to R \tag{9.4}$$

is monotone increasing.

One of the principal aims of the theory of sorting algorithms is to devise efficient algorithms for constructing such maps. We now describe a methodology for doing so.
Let

$$X = \{\phi: \{1, \ldots, k\} \to S\} \tag{9.5}$$

denote the set of maps with domain the first k integers and with image in S. The permutation group G of the set S acts as a transformation group on the space of maps (9.5) in the following way:

$$(g\phi)(j) = g(\phi(j)), \quad \begin{cases} \text{for } j = 1, \ldots, k, \\ \text{for } \phi \text{ a map of } \{1, \ldots, k\} \to S. \end{cases} \tag{9.6}$$

The sorting algorithms described by Knuth seem to be associated naturally with this action of G on X. Let us then formulate them in the following way:

DEFINITION. *A trajectory system for a sorting algorithm of the set S with a given key* κ is defined by the following data:

(a) A subset U of the permutation group G of S.
(b) A sequence $u_1, u_2, \ldots$ of elements of U such that the following condition is satisfied:

> For $\phi \in X$, $\phi_j = u_j(\phi)$, the sequence $\phi_1, \phi_2, \ldots$ ends after a finite number of steps at an element of X such that the composite map $\kappa\phi$ is monotone.

As for the methods for the GCD algorithms given in Section 4, our aim is to show how the commonly-used sorting algorithms may be described via the algebraic analogue of a closed-loop feedback structure.

Much of the algebraic structure of the permutation group may be thought of 'geometrically' as involving ways of relating the structure of subsets of the underlying sets and subgroups of the permutation group. Let us recall some of this material, since it plays an important role in the sorting algorithms.

Given a subset

$$Z \subset S, \tag{9.7}$$

let us define the following subsets of G:

$$G_Z = \{g \in G : g(Z) \subset Z\}, \tag{9.8}$$

$$G'_Z = \{g \in G : g(Z) \subset Z, \, g(S - Z) \subset S - Z\}. \tag{9.9}$$

The proof of the following result should be obvious.

THEOREM 9.1. *For each subset* $Z \subset S$, *the subsets* G_Z, G'_Z *of* G *defined by* (9.8) *and* (9.9) *are subgroups of* G, *with* $G_{Z'}$ *contained in* G_Z.

The algorithm called *bubble-sort* in the computer science literature involves sorting by choice of successive transpositions. Let us recall the following definition.

DEFINITION. Let (x, y) be an ordered pair of elements of S, with $x \neq y$. Define an element of G: $g_{(x,y)}: S \to S$ as follows:

$$g_{(x,y)}(z) = \begin{cases} z, & \text{if } z \in S - \{x, y\}, \\ y, & \text{if } z = x, \\ x, & \text{if } z = y. \end{cases} \tag{9.10}$$

The elements of the permutation group of the set S obtained in this way are called *transpositions*.

THEOREM 9.2. *Consider the map defined by (9.10):*

$$(x, y) \to g_{(x,y)}, \qquad S \times S - \{(x, x) : x \in S\} \to G. \tag{9.11}$$

Then, the map (9.11) is invariant under the two-element permutation:

$$S \times S \to S \times S, \qquad (x, y) \to (y, x) = \alpha(x, y). \tag{9.12}$$

In formulas, this means that

$$g_{\alpha(x,y)} = g_{(x,y)}. \tag{9.13}$$

Let S^* be the orbit space of $S \times S - \{(x, x) : x \in S\}$ under the cyclic transformation group of order 2 defined by (9.13). More explicitly, this means that for $x, y, x', y' \in S$:

$$(x, y) \sim (x', y'),$$

if and only if $x \neq y$, $x = y'$ and $y' = x'$. $\tag{9.14}$

S^* is the quotient of $S \times S - \{(x, x)\}$ under this equivalence relation. Then, the map (9.11) passes to the quotient to define a map

$$S^* \to G. \tag{9.15}$$

Proof. Notice that the definition (9.10) is invariant if one permutes x and y. This proves (9.19). The existence of the map (9.15) follows from general set-theoretic facts. $\square$

Let:

$$U = \text{set of transposition elements of } G. \tag{9.16}$$

Then, the group action

$$G \times S \to S \tag{9.17}$$

together with the inclusion map

$$U \to G \tag{9.18}$$

defines a *control system*, with G as a *structure group* and *outputs = states*. This control system is the underlying one in the bubble-sort program, which we now consider in more detail.

Here is a result that follows from the standard result in the algebra of the symmetric group that an arbitrary element of G may be written as a product of transpositions.

THEOREM 9.9. *The control system (S, G, U) defined above is* controllable *in the sense that the following condition is satisfied:*

$$\text{Given points } x_0, x_1 \in S, \tag{9.19}$$

there is an open-loop control $j \to u(j) \in U$ such that the solution of the following difference equation

$$x(j + 1) = u(j + 1)(x(j)), \quad x(0) = x_0, \tag{9.20}$$

satisfies the following end-point condition: $x(n) = x_1$.

Let us now turn to a discussion of the *bubble sort* algorithm from this point of view. Suppose that the set S comes provided with two fixed structures:

(a) A *key map*

$$\kappa: S \to R \tag{9.21}$$

which is one-one;

(b) A map

$$\phi_0: \{1, \ldots, k\} \to S \tag{9.22}$$

which identifies S with the set of the first k positive integers.

We have defined U as the set of transposition elements of G. Using the key κ, we can now define a subset U_κ of U as follows:

U_κ consists of the transpositions associated with the elements

$$\{(\phi(j), \phi(j + 1)); j = 1, \ldots, k - 1\}$$

such that

$$\kappa(\phi(j + 1)) > \kappa(\phi(j)). \tag{9.23}$$

In other words, U consists of the transpositions that *improve* the desired monotonicity of the key function. The intuition is that the action of the elements U_κ serve to 'bubble up' the elements of the set S which have larger key values.

The *quick sort* algorithms involve the following *divide-and-conquer* ideas. Split up S into the union of two subsets

$$S = S_1 \cup S_2. \tag{9.24}$$

Set

$$\kappa_1 = \kappa \text{ restricted to } S_1, \tag{9.25}$$

$$\kappa_2 = \kappa \text{ restricted to } S_2. \tag{9.26}$$

We can now – 'recursively' – assume that we have available sorting algorithms for the keys κ_1 and κ_2. If we have chosen the decomposition (9.24) cleverly, these algorithms will combine to give one for the key κ. For details, refer to the standard references [2, 16, 23].

Lyapunov functions do not play an explicit role in the formulation of these algorithms. Perhaps one can put them within that framework by formulating a *distance function* on the space of sorting maps which measures in an appropriate way the distance from the set of maps whose composition with the key map is monotone. Since the underlying structure group G is *finite* such a metric can be chosen which is invariant under right and left translation.

10. The Recursion Relations of the Binomial Coefficients as an Example of a System of Partial Difference Equations and Solution as a Fixed Point

In preceding sections, we have developed a general geometric way of thinking about recursive programs. It will, of course be, useful to apply this way of thinking to as wide a variety of situations as possible. I believe that some of the most interesting examples will occur within the realm of 'partial' recursion relations. Special cases of this are partial difference equations. In the computer science literature, such mathematical objects are often treated as examples of parallel or concurrent processes.

In the paper by Manna and Pneueli [21] on the application of temporal logic to concurrent processes, the computation of the binomial coefficients

$$\binom{n}{m} = \frac{n!}{m!(n-m)!} \tag{10.1}$$

provides a main example. This example is also of interest from our more geometric viewpoint, since computing (10.1) via its recursion relations, amounts to solving a partial difference equation. In this section, we shall present this point of view.

Start off with (10.1) defined via its generating function by the formula

$$(1 + x)^n = \sum_{m=0}^{n} \binom{n}{m} x^m. \tag{10.2}$$

To find a recursion relation, differentiate both sides of (10.2) with respect to x:

$$n(1 + x)^{n-1} \equiv \sum_{m=1}^{n} \binom{n}{m} m x^{m-1}$$

$$= \sum_{m=0}^{n-1} \binom{n}{m+1} (m+1) x^m \tag{10.3}$$

$$= \sum_{m=0}^{n-1} n \binom{n-1}{m} x^m. \tag{10.4}$$

Equating coefficients of x^m on both sides of this equation, gives the following relations:

$$n \binom{n-1}{m} = (m+1) \binom{n}{m+1}, \qquad m \binom{n}{m} = n \binom{n-1}{m-1}. \tag{10.5}$$

We have proved the following theorem.

THEOREM 10.1. *Let* X *be the set*

$$X = \{(m, n) : m, n \in N, 0 \leqslant m \leqslant n\}. \tag{10.6}$$

Then

$$(m, n) \to \binom{n}{m}$$

is the unique map $X \to N$ *satisfying recursion relation* (10.5) *plus the boundary conditions*

$$\binom{n}{n} = 1 = \binom{n}{0}, \quad \text{for } n \in \mathbf{N}. \tag{10.7}$$

Proof. By induction on n. $\qquad\qquad\square$

Let us set up the solution of the partial difference equation (10.5) as a fixed point problem, in the style of Scott [19, 22, 24]. Let

$$\Gamma' = \{X \to' \mathbf{N}\} \tag{10.8}$$

be the space of partial maps from X to the integers. Define a map $\alpha : \Gamma' \to \Gamma'$ by the formula

$$\alpha(\gamma')(n, m) = \frac{n}{m} \, \gamma(n-1, m-1)$$

$$\text{if the right-hand side is defined, and if } n, m \geqslant 1. \tag{10.9}$$

$$= 1 \quad \text{if } n \text{ or } m < 1. \tag{10.10}$$

$$\text{Not defined if neither of the above}$$
$$\text{conditions is satisfied.} \tag{10.11}$$

THEOREM 10.2. *Let* $\gamma \in \Gamma'$ *be a fixed point of* α, *the map defined by* (10.9)–(10.11). *Then,* γ *is a total map:* $\Gamma' \to \mathbf{N}$, *and a solution of the partial difference equation* (10.5), *with boundary conditions* (10.7).

Proof. Should be clear. $\qquad\qquad\square$

Let us now construct a fixed point of α by iteration. Pick a point $\gamma_0 \in \Gamma'$, and set

$$\gamma_1 = \alpha(\gamma_0)$$

$$\vdots$$

$$\gamma_n = \alpha(\gamma_{n-1}) \tag{10.12}$$

$$\vdots$$

Start off with γ_0 as follows:

$$\gamma_0(m, n) = \begin{cases} 1, & \text{if } m = n = 0, \\ 1 \text{ is not defined if } m \text{ or } n > 0. \end{cases} \tag{10.13}$$

The reader will now check that when (10.12) is made explicit, it reproduces the computation of the binomial coefficients by iteration.

11. Triangularly Recursive Computational Processes

We can abstract a general view from the example of the binomial coefficients treated in Section 10. In Section 1, we have supposed that a computer program provides the following data:

$$\text{a sequence } n \to X_n \text{ of sets,} \tag{11.1}$$

$$\text{a collection of sequences } n \to X_n \in X_n \text{ of elements of } X_n. \tag{11.2}$$

Such a program will be called *recursive* if each data point x_n is defined in terms of those which 'temporarily' precede it.

So far, we have not specified the structure of the sequence of sets $\{X_n\}$. In the example of the computation of the binomial coefficients – and in many other situations in computer science – each X_n itself consists of sequences. Let us now formulate a general situation. Suppose that

$$x_n = \{x_m(k) : 0 \leqslant k \leqslant m(n), \, x_m(k) \in X_{n,k}\}, \tag{11.3}$$

where

$$\{X_{n,k} : 0 \leqslant k \leqslant n, \, n = 0, 1, \ldots\} \tag{11.4}$$

is a collection of sets depending in the 'triangular' way indicated in (11.4) on the pairs of integers (n, k).

In addition, let us suppose that x_n is determined by a first-order difference equation of the form

$$x_{n+1} = f_n(x_n). \tag{11.5}$$

and that the map $f_n : X_n \to X_{n+1}$ which occurs in (11.5) has the form

$$f_n(x_n) = F_n^1(x_n(1), \ldots, x_n(n)), \ldots, F_n^{n+1}(x_n(1), \ldots, x_n(n)). \tag{11.6}$$

The difference equation (11.5) then takes the form

$$x_{n+1}(m) = F_{n+1}^m(x_n(1), \ldots, x_n(n)), \quad \text{for } m = 1, \ldots, n+1. \tag{11.7}$$

We call such a system a *triangularly recursive* computational system.

12. The Geometry of Paths. Prolongation and Correspondence. Contact Transformations

One of the aims of this paper is to develop ways of describing and studying computer programs that are analogous to the geometric theory of path systems. One of the most important geometric ideas associated with such systems is that of equivalence, in the sense used in the classic work of Sophus Lie and Elie Cartan. In order to provide some intuition and motivation, we shall now briefly review how these notions are defined for the classical, continuous situation, then go in the following sections to define analogous ideas for the discrete, computer science situation.

Begin as follows. Let X be a finite dimensional, C^∞, paracompact manifold. Let R be the real numbers considered as such a manifold, parameterized by a 'time' variable 't'. Let $J^r(R, X)$ be the manifold of r-jets of smooth mappings $R \to X$.

Geometrically [12], $J^r(R, X)$ may be thought of as the space of rth *order tangent elements* to smooth curves in X.

DEFINITION. An rth *order path system* on X is defined by a smooth submanifold

$$\mathbf{P} \subset J^r(R, X). \tag{12.1}$$

A smooth curve

$$\mathbf{x}: R \to X, \quad t \to x(t)$$

is said to *belong to the system* $\mathbf{P}$ if the following condition is satisfied:

$$j^r(\mathbf{x})(R) \subset \mathbf{P}. \tag{12.2}$$

$j^r(\mathbf{x})$ is a map: $R \to J^r(R, X)$ which is canonically defined in the Ehresmann jet-space theory, called the *r-jet* or rth *prolongation* of $\mathbf{x}$.

DEFINITION. Let X and Y be manifolds with

$$\mathbf{P} \subset J^r(R, X), \qquad \mathbf{Q} \subset j^r(R, Y)$$

path systems on X and Y, respectively. Let $\phi: X \to Y$ be a smooth map. ϕ is said to define a *prolongation* of the path system $\mathbf{P}$ to the path system $\mathbf{Q}$ if the following condition is satisfied:

For each curve $\mathbf{x}$ in the path system $\mathbf{P}$, the image curve

$$\phi(\mathbf{x}): t \to \phi(x(t)) \tag{12.3}$$

lies in the path system Q.

Let us now extend this prolongation idea to cover what in the classical geometry literature is called a correspondence.

DEFINITION. Let X and X' be manifolds which carry path systems $\mathbf{P}$ and $\mathbf{P}'$. A *correspondence* from the system $(X, \mathbf{P})$ to the system $(X, \mathbf{P}')$ is defined by the following data:

A third path system $(Z, \mathbf{Q})$

and maps

$$\phi: Z \to X, \qquad \phi': Z \to X'$$

such that the following conditions are satisfied:

ϕ is a prolongation map from path system $(Z, \mathbf{Q})$ to path system $(X, \mathbf{P})$.

ϕ' is a prolongation map from path system $(Z, \mathbf{Q})$ to the path system $(X', \mathbf{P}')$.

To put geometric flesh on these bare definitional bones, let us review a main historical example – *contact transformations* — which, in Lie's hands [17], motivated the general ideas.

Contact transformations between path systems in the plane specialize the general definitions in the following way:

$X = R$. Label the coordinate on X as "x".

$r = 1$.

$J^1(R, X)$ is a manifold with a coordinate system labeled as (x, x', t).

$f: J^1(R, X) \to R$ is a smooth, real-valued function written in these local coordinates as $f(x, x', t)$.

$P \subset J^1(R, X)$ is the set of (x, x', t) such that $f(x, x', t) = 0$.

A smooth curve $t \to x(t)$ in X then belongs to the path system $\mathbf{P}$ if and only if it satisfies the following first-order ordinary differential equation

$$f\left(x, \frac{dx}{dt}, t\right) = 0. \tag{12.4}$$

Remark. We use the classical term 'in the plane' for this ordinary differential equation because geometrically the graph $t \to (t, x(t))$ of each solution determines a curve in the 'plane', i.e., in R^2.

Let us suppose that (Y, Q) is another path system determined by an ordinary differential equation of similar form on Y, another copy of the real line

$$g\left(y, \frac{dy}{dt}, t\right) = 0,$$

where (y, y', t) are coordinates on $J^1(R, Y)$.

Here is the classical definition.

DEFINITION. A *contact transformation* between systems is defined by a map $\phi: J^1(R, X) \to J^1(R, Y)$ satisfying the conditions

There are smooth functions $h, k: J^1(R, X) \to R$ such that

$$\phi^*(dg - g'\,dt) = h(dx - x'\,dt), \quad \phi^*(g) = kh. \tag{12.5}$$

Such a map then defines a map sending paths $\mathbf{x}: t \to x(t)$ in the path system $\mathbf{P}$ to curves $\mathbf{y}: t \to y(t)$ in the path system $\mathbf{Q}$, with the following geometric property:

> If $\mathbf{x}: t \to x(t)$, $\mathbf{z}: t \to z(t)$ are two paths in the path system $\mathbf{P}$ having the property that their graphs
>
> $$t \to (t, x(t)), \qquad t \to (t, z(t))$$

have first-order contact at one point, then their image curves under the path system morphism attached to ϕ also have first-order contact at one point.

This geometric property is, of course, what led Lie to give these transformations the 'contact' name.

13. Prolongations and Correspondences for Trajectories of Programs

Let us now carry over concepts of classical differential geometry to the trajectory systems associated with computer programs.

The basic geometric objects to be studied here are, as before, fiber spaces

$$\pi: Z \to T, \tag{13.1}$$

$$t \to X(t) \subset Z, \tag{13.2}$$

with base the ordered set T.

DEFINITION. Let

$$\pi: Z \to T, \qquad \pi': Z' \to T$$

be two such fiber spaces, with the same base space. A map

$$\phi: Z \to Z' \tag{13.3}$$

is said to define a *prolongation* between the two structures if the following condition is satisfied:

$$\pi\phi = \pi' \tag{13.4}$$

Geometrically, (13.4) means that ϕ maps each fiber $X(t) = \pi^{-1}(t) \subset Z$ into the fiber $X(t)' = \pi_n'^{-1}(t) \subset Z'$.

DEFINITION. Let

$$\pi: Z \to \mathbf{T} \qquad \pi': Z \to \mathbf{T}'$$

be two fiber spaces. A *correspondence* between them is defined by the following data:

(a) A third fiber space $\pi'': Z'' \to \mathbf{T}$;
(b) a pair $\phi: Z'' \to Z$, $\phi': Z'' \to Z'$ of maps which are prolongation maps in the fiber space sense, as defined above.

Remark. This basic geometric picture – a relation between geometric objects on two spaces via a structure on a third space and a pair of projection maps – also occurs in geometry in the theory of *birational correspondences* in algebraic geometry and the theory of *Bäcklund transformations* in differential geometry.

Now we can extend these notions to computer programs. Recall that if $\pi: Z \to \mathbf{T}$ is a fiber space and

$$\Gamma(\pi) = \{\gamma: t - \gamma(t) \in X(t) = \pi^{-1}(t)\}$$

is the space of cross-section maps, then the trajectories of an associated computer program define a subset $\Gamma \subset \Gamma(\pi)$.

Suppose now that

$$\pi: Z \to \mathbf{T}, \qquad \pi': Z \to \mathbf{T}', \qquad \pi'': Z \to \mathbf{T}''$$

are three such fiber spaces, with three associated computer program trajectories G, G', G''.

DEFINITION. A prolongation map $\phi: Z \to Z'$ for fiber spaces is also a prolongation map with respect to computer program trajectory systems on Z and Z' if the following condition is satisfied:

If $\gamma: t \to \gamma(t)$ belongs to Γ, then

$$\phi(\gamma): t \to \phi\gamma(t) \tag{13.5}$$

belongs to Γ'.

DEFINITION. A pair (ϕ, ϕ') of prolongation maps

$$\phi: Z'' \to Z, \qquad \phi': Z'' \to Z'$$

for the computer program trajectory systems (Γ'', Γ), (Γ'', Γ'), respectively, is said to define a *correspondence* between the pair of programs Γ, Γ'.

Remark. There is considerable informal discussion in the computer programming literature about 'transformations' between computer programs. I hope that the definitions presented above based on analogous concepts for differential and algebraic systems will provide some sort of foundation for precise mathematical discussion of these important concepts.

14. The Temporal Logic of Programs and the Theory of Correspondences in Set Theory

Mathematical logic enters at many points in computer science. (Indeed, the cliche that it is 'the calculus of computer science' is true!). For computer programming theory, it enters in (at least) three ways. First, it is the foundation for any reasonable theory of semantics of programs. Second, it is essential to any theory of knowledge representation, and is used there to organize reasoning about the objects described by the program. Finally, it can be used to reason about the properties of programs.

In all three of these applications of logic, the classical ideas must be supplemented by other mathematical insights. Traditionally, logic is concerned with statements about broad types of reasoning – Gödel's theorems being the most famous – not about how it can be used to achieve certain purposes. There is a field called computational logic which does address some of the issues necessary for such applications. However, I believe that in addition the algebraic approach to logic – via Boolean and Heyting algebras, categories and toposes, sheaves, etc. – has much to offer for any broad program of application, especially if its diffuse structure can be focussed in more concrete ways. Certainly, the experience over the past 30 years in control system theory is that algebrazation provides a useful formalization and sharpening of material, which can lead to further important insights.

Saying anything serious about the semantics and knowledge representation aspects of the applications of logic is beyond the immediate scope of this paper. What does fit in very nicely with the 'geometric' approach advocated here is the temporal logic approach to the description of computer programs, especially the form in which this branch of modal logic is used in a paper by Manna and Pneueli [21]. For further background information, see [10, 19]. To close this paper, I present some material designed to link up the 'geometric' approach and the temporal logic operators used by Manna and Pneueli. In fact, I will show that these operators again have the same geometric form as the *correspondences* in classical differential and algebraic geometry, applied above to the trajectory systems for programs.

Consider again a computer program that defines a family Γ of trajectories, considered as a subset of the space of cross-sections of a fiber space

$$\pi: Z \to T. \tag{14.1}$$

Think of the base space T as a set representing time. A point $z \in Z$ then has a representation as

$$z = (t, x(t)), \tag{14.2}$$

where $t = \pi(z)$ and $x(t)$ is a point in the fiber $X(t) = \pi^{-1}(t)$ of the fiber space (14.1) above the point t of the base. We can think of t as the time, $x(t)$ as the space coordinate of the point z.

Consider the set

$$\Gamma \times Z. \tag{14.3}$$

In discussing properties of the program that generates the trajectory system Γ, we often want to make statement of a logical nature that refer to the set (14.3). For example, we might be given a subset S of Z and trajectory γ of Γ and a $t \in T$ and want to know if $\gamma(t)$ belongs to S. Given the set-theoretic way of thinking about logic, we might consider the Boolean algebra

$$\mathbf{B}(G \times Z) \tag{14.4}$$

of subsets of $G \times Z$ as the basic algebraic object to study. (The Boolean operations are union, intersection and complementation.) We can then construct a 'language' associated with this algebra, built up out of statements like

$$(\gamma, t, x) \in S, \tag{14.5}$$

for $S \in \mathbf{B}(G \times Z)$. We shall rename the statement (14.5) in terms of the 'predicate calculus' language as:

$$P_S(\gamma, t, x) \tag{14.6}$$

and regard (14.6) as a 'predicate' having the 'value *true*' if (14.5) is satisfied, '*false*' otherwise. Mathematically, we can identify the 'algebra' of 'predicates' of form (14.6) with the Boolean algebra (14.4).

One strategy that computer scientists traditionally used to 'reason' about the properties of computer programs is – following the ideas of the Modal logicians – to add to the 'language' of the predicates like (14.6) 'new' objects, such as

$$\square P_S(g, t, x) \tag{14.7}$$

For example, in temporal logic (14.7) 'means' the following statement:

If '$\square P_S(g, t, x)$ is true' then $P_S(g, t', x)$ is true for all t' greater or equal to t.

In this context, $\square$ is called the *henceforth* operator.

Rather than follow the route that is traditional in modal logic and first constructing a 'language' out of objects like (14.7), and then defining a semantics as – in the standard Kripke theory – as a set-theoretic interpretation, I will follow Jonsson and Tarski [15] to define objects like $\square$ directly as mappings on Boolean algebras.

Now, one natural mathematical way of defining mapping on Boolean algebras is to realize – via the Stone representation theorem, for example – the Boolean algebra as a subalgebra of the algebra of all subsets of a set S, construct a mapping $\phi: S \to S$, and map a subset A of S into $\phi(A)$ or $\phi^{-1}(A)$. We shall show that the simplest modal operators of temporal logic are the composition of two such 'geometric' mappings on sets. Again, this will have the same algebraic structure as the classical notion of *correspondence* in algebraic geometry and *Bäcklund transformation* in differential geometry.

In the next section, let us present the material we need in generality.

15. Boolean Properties of Mappings Between Sets

THEOREM 15.1. *Let* $\phi: S \to S'$ *be a mapping from the set S to the set S'. Let A', B' be subsets of S' and A, B be subsets of S. Then the following relations hold:*

$$\phi^{-1}(A' \cap B') = \phi^{-1}(A') \cap \phi^{-1}(B'), \tag{15.1}$$

$$\phi^{-1}(A' \cup B') = \phi^{-1}(A') \cup \phi^{-1}(B'), \tag{15.2}$$

$$\phi(A \cap B) \subset \phi(A) \cap \phi(B), \tag{15.3}$$

$$\phi(A \cup B) = \phi(A) \cup \phi(B). \tag{15.4}$$

Proof. Suppose that

$$a' \in A' \quad \text{and/or} \quad B'.$$

Then, $\phi^{-1}(a') \in \phi^{-1}(A') \cap/\cup \phi^{-1}(B')$.

Conversely, suppose that

$$a \in \phi^{-1}(A') \cap/\cup \phi^{-1}(B').$$

Then, $\phi(a) \in A' \cap/\cup \phi^{-1}(B')$. This proves (15.1)–(15.2).

Suppose now that

$$a \in A \quad \text{and/or} \quad B.$$

Then, $\phi(a) \in \phi(A)$ and/or $\phi(B)$, whence (15.3) and the relation

$$\phi(A \cup B) \subset \phi(A) \cup \phi(B). \tag{15.5}$$

To finish the proof of (15.4), suppose that $a \in \phi(A)$ or $\phi(B)$. For example, suppose that $a' = \phi(a)$, with $a \in A$. Then, $a' \in \phi(A \cup B)$.

The other possibility similarly leads to the relation

$$\phi(A) \cup \phi(B) \subset \phi(A \cup B) \tag{15.6}$$

which completes the proof of Theorem 15.1.

We can now set up correspondences by means of a pair

$$\phi': S \to S', \qquad \phi'': S \to S'' \tag{15.7}$$

of maps between sets.

DEFINITION. Let S' and S'' be a pair of sets, with $\mathbf{B}(S')$ and $\mathbf{B}(S'')$ the Boolean algebras of their subsets. A map

$$\phi: \mathbf{B}(S') \to \mathbf{B}(S'') \tag{15.8}$$

is said to be a *modal operator* associated with the pair (ϕ', ϕ'') of maps indicated in (15.7) if ϕ is defined in terms of (ϕ', ϕ'') by the formula

$$\phi = \phi''\phi'^{-1}. \tag{15.9}$$

THEOREM 15.2. *Suppose that* $\phi: \mathbf{B}(S') \to \mathbf{B}(S'')$ *is a modal operator, in the above sense. Then, f satisfies the following relations vis-à-vis the Boolean operations*

$$\phi(A' \cap B') \subset \phi(A') \cap \phi(B'), \tag{15.10}$$

$$\phi(A' \cup B') = \phi(A') \cup \phi(B') \tag{15.11}$$

for $A', B' \in \mathbf{B}(S')$.

Proof. Follows from (15.1)–(15.4) and (15.9). $\qquad\square$

16. The 'Henceforth' Operation of Temporal Logic

Let us now see how the set-theoretic operation ϕ defined above works for the basic operation of temporal logic [10, 20]. Let T be a set with a partial order relation

$$(t_1, t_2) \to t_1 \leqslant t_2. \tag{16.1}$$

Suppose that

$$\mathbf{P}: t \to P(t) \tag{16.2}$$

is a map from T to a Boolean algebra $\mathbf{B}$.

In Temporal Logic, $\Box\mathbf{P}: t \to \Box P(t)$ is supposed to be the map with the following property:

If $\Box P(t)$ is true in a given valuation, then
$P(t')$ is true in this valuation for $t' \geqslant t$. $\tag{16.3}$

Let us construct a set-theoretic model for such an operation. Let S be a set. Suppose that

$$\mathbf{B} = \mathbf{B}(S). \tag{16.4}$$

We assign the following subset of $S \times T$ to $\mathbf{P}$:

$$\mathbf{P} = \{(s, t) \in S \times T : s \in P(t)\} \tag{16.5}$$

Assign the following 'predicate' function $(s, t) \to P(s, t)$ to $\mathbf{P}$:

$P(s, t)$ is *true* if and only if $s \in P(t)$. $\tag{16.6}$

Now, set

$$Z = \{(s, t, t') : s \in S, t, t' \in T; t' \geqslant t\}. \tag{16.7}$$

The following maps

$$\phi': Z \to S \times T, \tag{16.8}$$

$$\phi'': Z \to S \times T \tag{16.9}$$

are defined as

$$\phi'(s, t, t') = (s, t), \tag{16.10}$$

$$\phi''(s, t, t') = (s, t'). \tag{16.11}$$

Define the map

$$\phi: (\text{subsets of } S \times T) \to (\text{subsets of } S \times T) \tag{16.12}$$

as

For $P \subset S \times T$, $\phi(P) =$ the set of (s, t') such that there is a $t \in T$ such that

$$t \leqslant t' \tag{16.13}$$

and

$$(s, t) \in P. \tag{16.14}$$

THEOREM 16.1.

$$\phi(P) = \phi'' \phi'^{-1} \tag{16.15}$$

Proof. Set

$$P' = \phi'' \phi'^{-1}(P) \tag{16.16}$$

Then, $(s, t') \in P'$ if and only if there is a t such that the following conditions are satisfied:

$$t \leqslant t', \quad (s, t, t') \in \phi'^{-1}(P),$$

i.e. $(s, t) \in P$.

This proves 6.15. $\qquad\qquad\square$

Convert $\phi(P)$ to a predicate function $(s, t') \to \phi(P)(s, t')$ following (16.6), then, we have

$\phi(P)(s, t')$ is true if there is a $t \in T$ such that

$$t \leqslant t', \tag{16.17}$$

$$P(s, t) \text{ is true.} \tag{16.18}$$

This verifies that

$$\phi(P) \equiv \square P \tag{16.19}$$

is the 'henceforth' operator of temporal logic. In turn, this set-theoretic model is essentially the well-known Kripke semantics for the modal operator $\square$.

Acknowledgement

I would like to thank Gerald Sussman, with whom I have had many useful discussions about the relation between mathematics, control and computer science and who has generously given me tutorials in the theory and practice of hacking and programming.

References

1. Abelson, H. and Sussman, G. J.: *Structure and Interpretation of Computer Programs*, MIT Press, Cambridge, Mass., 1985.
2. Aho, A., Hopcroft, J., and Ullman, J.: *Data Structures and Algorithms*, Addison-Wesley, Reading, Mass., 1983.
3. Alagic, S. and Arbib, M. A.: *The Design of Well-Constructed and Correct Programs*, Springer-Verlag, New York, 1978.
4. Anderson, B. and Moore, R.: *Linear Optimal Control*, Prentice-Hall, Englewood Cliffs, NJ. 1971.
5. Bell, J. and Machover, M.: *A Course in Mathematical Logic*, North-Holland, Amsterdam, 1977.
6. Caines, P. E.: *Linear Stochastic Systems*, Wiley, New York, 1988.
7. Ehresmann, C.: *Categories et structures*, Dunod, Paris, 1965.
8. Eilenberg, S. and Elgot, C.: *Recursiveness*, Academic Press, New York 1970.

182   ROBERT HERMANN

9. Floyd, R. W.: Assigning meanings to programs, in J. T. Schwartz, (ed.), *Proc. Symp. Appl. Math.* vol. 19, Amer. Math. Soc. Providence, R.I., 1967, pp. 19–32.

10. Goldblatt, R.: *Logic of Time and Computation*, Center for Study of Language and Information, Palo Alto, 1987.

11. Goldblatt, R. *Topoi*, North-Holland, Amsterdam, 1984.

12. Hermann, R.: *Geometry, Physics, and Systems*, M. Dekker, New York 1973.

13. Hoard, C. A. R.: An axiomatic basis for computer programming, *Comm. ACM* **12**, (1969) 576–580.

14. Johnstone, P. T.: *Stone Spaces*, Cambridge University Press, 1982.

15. Jonsson, B. and Tarski, A.: Boolean algebras with operators, I., *Amer. J. Math.* 1952.

16. Knuth, D.: *The Art of Computer Programming*, Vols. 1–3, Addison-Wesley, Reading, Mass.

17. Lie, S.: Geometrie der Beruhrungstransformationen, Leipzig, 1896. Reprinted by Chelsea, New York, 1977.

18. MacLane, S.: *Categories for the Working Mathematician*, Springer-Verlag, New York, 1976.

19. Manes, E. G. and Arbib, M. A.: *Algebraic Approaches to Program Semantics*, Springer-Verlag, New York, 1986.

20. Manna, Z.: *Lectures on the Logic of Computer Programming*, Society for Industrial and Applied Mathematics, Philadelphia, 1980.

21. Manna, Z. and Pneueli, A.: Verification of concurrent programs: the temporal framework, in R. S. Boyer and J S. Moore (eds.), *The Correctness Problem in Computer Science*, Academic Press, New York, 1981.

22. Scott, D. S.: Lectures on a mathematical theory of computation, in M. Broy and G. Schmidt, (eds.), *Theoretical Foundations of Programming Methodology*, D. Reidel, Dordrecht, 1982.

23. Sedgewick, R.: *Algorithms*, Addison-Wesley, Reading, Mass., 1983.

24. Stoy, J. E.: *Denotational Semantics: The Scott-Strachey Approach to Programming Languages*, MIT Press, Cambridge, Mass., 1977.

25. Wirth, N.: *Algorithms + Data Structures = Programs*, Prentice-Hall, Englewood Cliffs, NJ, 1976.

LECTURE NOTES FOR TALKS ON GEOMETRIC COMPUTING SCIENCE

To further orient the reader about my goals, Here are the notes for lectures I gave in the Spring of 1990 on this material.

GEOMETRIC ANALOGIES BETWEEN THE THEORY OF COMPUTATION AND CONTROL THEORY

ABSTRACT

This talk presents idea from three recent papers: A 'GEOMETRIC' VIEW OF THE DYNAMICS OF TRAJECTORIES OF COMPUTER PROGRAMS, A CATEGORY THEORY-BASED GENERALIZED DIFFERENTIAL SYSTEM FORMALISM FOR USE IN COMPUTER SCIENCE, NUMERICAL ANALYSIS AND CONTROL THEORY and PATH GEOMETRY ANALOGIES BETWEEN CONTROL THEORY AND THE THEORY OF COMPUTATION. The common theme is the adaptation to Computer Science purposes of the 'geometric' ways of thinking that have been useful in areas of pure and applied mathematics (e.g. physics, mechanics, control theory and logic) to the computational side of computer science. First, the common mathematical structure of control and automata theory will be reviewed. Then it will be shown how some simple textbook algorithms (e.g. GCD and the Sorting Algorithms) can be viewed as simple feedback control systems, somewhat analogous to the Linear Regulator, i..e the system dual to the 'Kalman Filter'. Second, the one-step numerical algorithms of the numerical analysis of ODE's will be interpreted in a Generalized Lie-Theory framework, involving, algebraically, certain types of categories and functors and, analytically, certain types of topological structures on these categories. Finally, ideas of Path System Geometry will be generalized to cover some Deduction Systems of computational Logic, emphasizing the anology between the concepts of 'controllability', 'the reachable set' and ' Godel completeness'.

'Geometric' approach to organizing CS

(Parallel to geometric approaches to physics, mechanics, control, logic. Involves analogue of 'manifolds', 'fiber bundles', 'differential operators', jet-bundles and mappings, etc.)

'APPLIED' MOTIVATION (NASA SPACE STATION, ETC.)

MIX TRADITIONAL CONTROLS (USING 'REAL NUMBERS') AND LANGUAGE, LOGIC, ETC.

METHODS OF DEALING WITH ENGINEERING CONTROL PROBLEMS IN WHICH LANGUAGE, LOGIC, ETC. ARE INGREDIENT

STUDY 'DYNAMIC' ASPECTS OF CS/AI PROGRAMS

SUPERVISORY CONTROL OF TRADITIONAL CONTROL SYSTEMS

COMBINE RECURSIVE FUNCTIONS (LISP, ETC.) AND MATHEMATICAL STRUCTURES THAT APPEAR IN DIFFERENTIAL GEOMETRY, MECHANICS, CONTROL, ETC. (WORK OF BLUM, SHUB, SMALE INVOLVES A 'DYNAMICAL SYSTEM THEORIST'S APPROACH)

Ways of 'structuring' data:

(1).PHYSICS AND MECHANICS

Dynamical systems

R^n, differentiable manifolds, Hilbert Spaces.

Configuration space, phase space, projective space of Hilbert Space.

(2). CONTROL THEORY

Input-output maps, 'machines', automata, state-space models, etc. When reformulated in terms of 'path systems' becomes very close to Computer Science.

(3) COMPUTER SCIENCE (Programming theory, recursive structures, Knowledge Representation, Logic-Based AI.)

(4) DIFFERENTIAL SYSTEMS (Jet bundles, exterior differential systems, etc.

GOAL: DESCRIBE AND ANALYZE MATHEMATICAL STRUCTURES INVOLVED IN COMPUTER SCIENCE

TYPICAL EXAMPLES:

a).Computation of n! - Discrete Dynamical System
,

b) Greatest Common Divisor - Feedback Control System, analogous to Linear Regulator

c) One-Step Numerical Approximation to ODE's.

d) Logical Deductiion Systems

Algebraic formulation of Discrete Path System

Let X be a set. X^* free monoid generated by X.

$$X^* = \{\gamma = x_0 .. x_n : x_0 \ldots , x_n \in X, n \in N\}$$

A **discrete path system** on X is a subset Γ of X^* which is fixed point set of mapping $\alpha: X^* \longrightarrow X^*$ continuous w.r. to certain topologies on X^*. (Dana Scott's work)

$$\Gamma = \{\gamma : \alpha(\gamma) = \gamma\}$$

Examples:

Discete Dynamical System:

$$x(n+1) = f(x(n))$$

$$\alpha(x_0 \ldots x_n) = (x_0, f(x_0) \ldots f(x_{n-1}))$$

Example: $x(n) = n!$

Feedback control systems

> **Open loop:**
>
> $$x(n+1) = f(x(n), u(n)) \ , \ u(n) \ \varepsilon \ U$$

Fixed point definition:

> $$\alpha((x_0, u_0)\ldots (x_n, u_n) = ((x_0,u_0) \ (f(x_0), u_1)\ldots (f(x_{n-1}), u_n)$$

Closed loop:

> $$x(n+1) = f(x(n), Kx(n)),$$
>
> $$K: X \dashrightarrow U$$

'Curry' the relation $x \dashrightarrow f(x, u)$ to express in terms of semigroups acting on X, automata, 'languages', etc.

Example: GCD computation

$$X = N \times N. = \{(m, n)\}$$

$$U = 2 \times 2 \text{ integer matrices}$$

$$u = (a,b,c,d), \ x = (m, n), \ f(x, u) = (am + bn, cm + dn) = u \cdot x$$

Choose $K(x)$ so that;

$$GCD(x) = GCD(K(x))$$

$$min(K(x)) < min(x)$$

$x \dashrightarrow GCD(x)$ is 'conserved quantity'.

$x \dashrightarrow min(x)$ is 'Liapounov function' or 'energy'

ONE-STEP NUMERICAL APPROXIMATION OF ODE

$dx/dt = f(x)$, $x \in R^n$

$x(n+1, h) - x(n, h) = hF(x(n, h),h)$

$F: (x, h) \to F(x, h)$

$F(x, 0) = f(x)$ ("Consistency')

Approximation of orbits of ODE by orbits of DIFFERENCE EQ.

$$\lim_{n \to \infty} y(n, t/n) = x(t)$$

Translate into statements in 'language' of topological structures on $Rn*$

Logic deductive systems

X = Propositional Algebra

$$\cup: X \times X \dashrightarrow X$$

$$\ulcorner: X \dashrightarrow X$$

$V(X, \cup, \ulcorner) = \{v: X \dashrightarrow \textbf{BOOL}: v \text{ is algebra}$ homomorphism $\}$

$$\gamma \varepsilon \Gamma \subset X^*, \quad V \subset V(X, \cup, \ulcorner)$$

Γ is monotone w.r. V if $v(\gamma(n)) = \textbf{T}$ all $n \varepsilon N$, $v \varepsilon V$

$$j^k(\gamma)(n) = (\gamma(n-k), \gamma(n-k+1), ..., \gamma(n)$$

'k-jet of γ', $\quad j^k(\gamma): N \dashrightarrow X^k$

$DE \subset X^{k.}$. γ is solution of the difference equation DE if

$$j^k(\gamma)(N) \subset DE$$

$\Gamma(DE) = \{\gamma \ \varepsilon \ X^*: \gamma \text{ is solution of DE}\}$

Translated to Logic: Γ is generated by recursive deduction rules:

REACHABLE SETS FOR PATH SYSTEMS

$\gamma \ \varepsilon \ \Gamma \subset X^*.$

$$R(x_0) = \{x_1: \text{there is } \gamma \ \varepsilon \ \Gamma \text{ such that}$$
$$x_0 = \text{begin}(\gamma), \ x_1 = \text{end}(\gamma)\}$$

$R(x_0) = $ **reachable set** from x_0 along path system Γ

Discuss relations to:

Controllablity and Stabilizability in Control Theory

Principle of "The reachable set consists of the points that are   not- unreachable"

"Completeness" in the sense of Logic (Godel and All That).

FIRST STEPS TOWARDS A DEFORMATION THEORY FOR PROCESSES

What is a Process? Informally, a description in terms of Logic (e.g. a Formal Language) of a discrete -or- continuous time 'dynamical system'.

'dynamical system' is taken in broad sense, to include 'control', Petri Net type things, 'operating systems', etc.

'Deformation Theory' - some way of considering families of Process depending on parameters.

Examples

Numerical Analyis: Difference Equations depending on parameters 'approximate' differential equations.

Physics: Relativistic physics, parameter is: Velocity of Light.

Quantum Mechanics: Parameter is: Planck's Constant.

Harmonic Analysis on groups:
 Fourier Series 'Approximates' Fourier Integral.
Algebraically, the Hopf algebra of functions on the additive group of the Real Line and the additive group of the Integers.

Computer-Science/Logic questions: How to use discrete/recursive structures to describe 'continuous' behavior?

EXAMPLE:

THE FORMULA FOR APPROXIMATION OF THE EXPONENTIAL FUNCTION; THE 'EULER' METHOD FOR ONE-STEP APPROX. OF LINEAR ODE'S.

to prove:

$$\exp(At) = \lim_{n \to} [1 + A/n]^n. \quad \text{'Lie-theoretically'}$$

A an mxm real matrix, an element of Lie algebra of
GL(m, R)

Theorem Let $\varepsilon \longrightarrow B(\varepsilon)$ be smooth curve in GL(m, R) such that:

$B(0) = 1$, the identity element.

$[dB/d\varepsilon](0) = A$.

Then,

$$\lim_{n \to} [B(t/n)]^n = \exp(tA)$$

In particular, can choose

$$B(\varepsilon) = 1 + A\varepsilon$$

Proof.

$$\log[B(t/n)^n] \; = \; n\log[B(t/n)] \; = \; t\log[B(t/n)]/[t/n]$$

$$\lim_{\varepsilon \,-\!\!->\, 0} \; \log[B(\varepsilon)]/\varepsilon \; = \; A \quad \text{(L'Hopital's Rule)}$$

Hence,

$$\lim_{n \,-\!\!->\, \infty} \; \log[B(t/n)]/[t/n] \; = \; A$$

$$\lim_{n \,-\!\!->\, \infty} \; [B(t/n)]^n \; = \; \exp(tA)$$

Deformation of ε-dependent family of one-parameter semigroups

$$n \; -\!\!-> \; [B(\varepsilon)]^n$$

into a one-parameter group

$$t \; -\!\!-> \; \exp(tA)$$

Algebraic model used in this work:

A Process is a functor between two categories.

Reference for Category Theory: Barr and Wells, Category Theory for Computing Scientists, Prentice-Hall.

Requires a Deformation Theory of Algebraic Structures Called 'Categories'. Modelled after "Deformation Theory' of Groups, Algebras, etc.

category C = {objects, morphisms}

    algebraic structure is partially defined map

        morphisms x morphisms --> morphisms

satisfying something like the 'associative law'.

Product defined by a subset of:

    π $\subset$ morphisms x morphisms x morphisms

Deformation is a family

      ε --> π^{ε} $\subset$ morphisms x morphisms x morphisms

of subsets satisfying 'algebraic conditions' required by Associative Law.

How to define a 'deformation theory'. Requires definition of a 'tangent space' - Algebraic geometry? Sheaf of rings defined on morphism spaces and their Cartesian products.

FIRST STEPS TOWARDS A DEFORMATION THEORY FOR PROCESSES

What is a Process? Informally, a description in terms of Logic (e.g. a Formal Language) of a discrete -or- continuous time 'dynamical system'.

'dynamical system' is taken in broad sense, to include 'control', Petri Net type things, 'operating systems', etc.

According to Dana Scott, the original "Process Theory' done in England was developed with 'operating systems' in mind.

'Deformation Theory' - some way of considering families of Process depending on parameters.

Examples

Numerical Analyis: Difference Equations depending on parameters 'approximate' differential equations.

Physics: Relativistic physics, parameter is: Velocity of Light.

Quantum Mechanics: Parameter is: Planck's Constant.

Harmonic Analysis on groups:
 Fourier Series 'Approximates' Fourier Integral.
Algebraically, the Hopf algebra of functions on the additive group of the Real Line and the additive group of the Integers.

Computer-Science/Logic questions: How to use discrete/recursive structures to describe 'continuous' behavior?

A NEW ALGEBRAIC SYMBOLIC- COMPUTATIONAL TOOL FOR CONTROL THEORY AND MECHANICS

Grossman, Gardner and Shadwick have been developing algebraic computational tools for Control based on the mathematics of Differential Geometry: Differential Forms, Vector Fields, Lie Brackets And All That. This talk describes a new computer language, called **CARTAN**, which is specialized to these computations. It is constructed by analogy with the Propositional Calculus of Logic, which is implemented in terms of either LISP or PROLOG. **CARTAN**

A NEW ALGEBRAIC SYMBOLIC- COMPUTATIONAL TOOL FOR NON-LINEAR CONTROL THEORY

Nonlinear feedback control   system

$$dx/dt = f(x, u)$$

Two ways of computing:

Vector fields and Lie Brackets

$$V^u = f(x, u)\partial /\partial x$$

 Lie, Hermes, Sussmann, Brocket, Krener, Hunt-Meyer-Su,

Differential forms, exterior derivative and multiplication:

$$\theta = dx - f(x, u) dt$$

(Cartan, Hermann, Gardner, Shadwick, ...)

Problems of mechanics (e.g. twin-lift helicopter, robots, space-craft have similiar mathematical form.)

Can use such systems as MACSYMA, but there seem to be inherent limitations.

Needed: Methods of symbolic computation with properties:

a)Avoid intermediate expression build-up

b). Have available properties of Functional Programming languages, such typing, recursion, abstraction, manipulate programs which contain programs, etc.

c) Facilitate conversion of symbolic output into Language needed for Numerical Computation:

Example: Conversion of MACSYMA output into FORTRAN code.

Sussman-HP are designing custom chips to do very fast, programmable ODE calculations.

Talks at Conference of Symbolic and Numerical Computation,

Sponsored by GE and SUNY-Albany, July, 1990. J. Mundy, Hopcroft, Sussman, Davenport, Zipplel, ...

Work done is still very primitive: Project for Ames?

Go back to first principles;consider problem of Logic and Logic Programming

The Language of Propositional Calculus.

Standard topic in Logic; also important as a simple prototype of many computer languages. . Start off with alphabet A consisting of the following objects:

Elements of $P, \cap, \cup, \neg, (,)$

P is a set; The elements of P are called the **atomic propositions**. $\cap$ and $\cup$ are called the **disjunction** and **conjunction** symbols. $\neg$ is the **negation** symbol,

Let A* be the strings in this alphabet. The Propositional Calculus is a Language based on the alphabet A i.e. a subset **P(A)**of A* defined by certain recursion rules, the so-called **well-formed formulas** of Logic. The following recursion rules determine the elements of**P(A)**:

The elements of P are well-formed formulas $\qquad$ (1)

If F, F' are well-formed formulas, then so are:

$$(F \cap F'), (F \cup F') \text{ and } \neg(F) \qquad (2)$$

P(A) is now defined as the smallest subset of A* satisfying (1)-(2).

Mathematically, PROLOG is a language customized to do calculations

in $A*$ and $P(A)$.

The language $P(A)$ of Propositional Logic may be considered

as 'free' version of the algebraic structures called Boolean

Algebras, which can - by a basic theorem of Marshall Stone- be

realized as Boolean algebras of subsets. There is a mapping of

$P(A\ )$ into a Boolean algebra called the **Lindenbaum- Tarski**

algebra , which - combined with Stone's Theorem - leads to what

logicians call a **Model of the Language** $P(A\ )$. Lindenbaum-

Tarski algebra is obtained from $P(A\ )$ by introducing relations.

Mathematical core PROLOG is to build sytem for keeping track of

these relations.

Proposed System :

Symbolic Computation in Control and Mechanics

Based Mathematically on Grassmann Algebras

as Propositional Calculus based on Boolean Algebras.

Differential Form Algebras: Grassmann Algebras
with exterior differentiation and Lie derivative
operators analogous to the 'modal' operators of Logic.

Start Alphabet C:

$$C = \{f, f_1, ..., \theta, \theta_1, ..., d, +, \wedge, (,)\}.$$

f is the symbol for 'functions', or '0-forms';

θ the symbol for 1-forms;

'd' the symbol for 'exterior derivative';

$\wedge$ the symbol for 'exterior multiplication'.

C stands for 'Elie Cartan',

A Language **L**(C) constructed called: **CARTAN**.

Models Associated to the Language **L**(C).

X= differentiable manifold.

involves almost the same 'alphabet' as the Propositional Calculus but it is 'typed' differently. Mathematically, it is a kind of Universal Algebra associated with Grassmann Algebra. Working out the details will involve many interesting and informative problems involving basic ideas of Computer Science and Mathematical Logic in a context that is much more directly applicable and relevant to engineering practice.

Integer r, $\mathbf{D}^r(X)$ = set of smooth differential forms of degree r on X.

'f' is associated with an element of $\mathbf{D}^0(X)$,

'θ' with an element of $\mathbf{D}^1(X)$,

'd' : $\mathbf{D}^r(X)$ --> $\mathbf{D}^{r+1}(X)$, exterior derivative operation

'$\wedge$' $\mathbf{D}^r(X)$ x $\mathbf{D}^s(X)$ --> $\mathbf{D}^{r+s}(X)$, exterior multiplication operation

'Knowledge Engineers' translate information about mechanical and control systems in terms of algebraic structure $\{\mathbf{D}^r(X), d, \wedge\}$, and then in terms of the Language $\mathbf{L}(C)$.

Involves process of **'Inverse Semantics'**

Final step: Translate' formulas Language $\mathbf{L}(C)$ into formulas in a known Programming Language such as PROLOG or LISP, or one of the dialects or specialized sub-language of LISP like MACSYMA..

Basic Data Structure   theory:

Set of sequences $(\theta_1, ..., \theta_m)$ of elements of $\mathbf{D}^1(X)$.

Mechanics: Data Structure sequences :

$(\alpha_1, ..., \alpha_n, \beta_1, ..., \beta_n)$

with the 2-form:

$$\omega = \Sigma \alpha_i \wedge \beta^i$$

defining the 'symplectic structure'

Mechanical system of n degrees of freedom

configuration coordinates $q = \{q^i: 1 \leq i \leq n\}$

momentum coordinates $p = \{p_i\}$.

$H(p, q) = $ **Hamiltonian** or **Total Energy**.

$$H_i = \partial H/ \partial q^i, \quad H^i = \partial H/ \partial p_i,$$

differential equations for the trajectory curves :

$$dq^i/dt = H^i, \quad dp_i/dt = -H_i.$$

Introduce differential forms:

$$\theta = \Sigma \, p_i dq^i - Hdt$$

$$\omega = d\theta = \Sigma \, dp_i \wedge dq^i - dH \wedge dt.$$

$$dH \wedge dt = \Sigma(H^i dp_i + H_i dq^i) \wedge dt$$

$$\omega = \Sigma(dp_i + H_i \wedge dt) \wedge (dq^i - H^i \wedge dt).$$

Set:
$$\alpha_i = dp_i + H_i \wedge dt,$$

$$\beta^i = (dq^i - H^i \wedge dt.$$

Solution curves of the Hamilton equations are curves on which the 1-forms α_i, β^i are zero,

Identity

$$\omega = \Sigma \alpha_i \wedge \beta^i$$

expresses the **symplectic structure** associated with the Hamilton equations

Passing from the Hamiltonian (i.e. total energy) function

H to the trajectory equations can be effected

algebraically by the factorization of the 2-form ω as a

'scalar product' (with respect to the Grassmann

algebra) of the n-vectors

$\alpha = (\alpha_1, ..., \alpha_n) , \beta = (\beta_1, ..., \beta_n)$

Feedback Control

$$dx/dt = f(x, u), \ x \ \varepsilon \ R^n, \ u \ \varepsilon \ R^m$$

$$\theta^1 = dx^1 - f^1 dt, \ \cdots, \ \theta^n = dx^n - f^n dt$$

Properties of control system translate into properties of

$$(\theta^1, \cdots, \theta^n)$$

Can be computed in terms of the Language $L(C)$ and the 'Inverse Semantics'.

Key calculation: **derived system**, defined as the collection of 1-forms θ such that:

$$\theta \wedge \theta^1 \wedge \ldots \wedge \theta^n = 0.$$

This calculation plays a key role in the theory of control via feedback linearization for nonlinear control systems.

Mathematics of 'feedback linearization' and constuction of 'control laws' can also be described in terms of this Grassmann algebra. .

Common engineering method of design of control:

Utilize a linear system of the form:

$$dz/dt = Az + Bv, \ z \ \varepsilon \ R^n, \ v \ \varepsilon \ R^m$$

as 'model'.

Consider following vector-valued 1- form on $R^{2n + 2m +1+ n \times n}$:

$$\mu = dz - (Az + Bv)dt - \Lambda(dx - f(x, u)dt$$

(The coefficients of the n x n matrix Λ are also variables.)

Finding **feedback linearization** of nonlinear system equiv. to:

Finding n-dimensional submanifolds of $R^{2n + 2m +1+ n \times n}$

on which the forms μ are zero.

Utilizing the algorithms introduced for doing this exactly or pertubatively involves - at the first stage - computing the 'derived systems' of the system of 1 - forms which are the components of μ.

Computation is formulatable in terms of Grassmann algebra.

ANALOGIES BETWEEN GODELIAN COMPLETENESS AND CARATHEODORY-CHOW CONTROLLABILITY.

ABSTRACT

This Chapter is concerned with collections of sequences of 'data', and mathematical schemes for giving structure to these sequences. As mathematical analogues, consider the role of sheaves and categories in giving a 'geometric' structure to Logic and the similiar role of differential system and Lie theory in the theory of Control. Analogies may be drawn between 'controllability' and ' completeness' (in the sense of Godel) for logical 'spaces' and 'systems'.   Following the work of Rasiowa and Sikorski, the basic 'geometric' structure is something like a 'fiber bundle' or 'foliated manifold', whose base space carries a Boolean algebra structure. The discrete analogs of Path Sytems are introduced algebraically as subsets of free monoids, and the iteration of the free-monoid construction provides an algebraic foundation for the theory of concurrent and parallel processes.   Path Systems, in the category of differentiable manifolds are the 'continuous' analogues of the 'discrete' structures encountered in various parts (Knowledge Representation, Programming Theory, Artificial Intelligence, etc.) of Logic-Based Computer Science.

1. INTRODUCTION.

The mathematical starting point for this paper is the geometric theory of **path systems.** In my paper [8] (reprinted in this Volume), I have suggested that 'discrete' versions of these geometric objects (plus associated 'fiber bundle' notions) could serve as a mathematical structure to organize some of the material to be found in the literature of Computer Programming Theory. In the present paper, I study geometric properties of such systems linked to what in the Control context is called 'controllability' and 'the reachable set'. I especially hope to develop analogous ways of studying what logicians call **completeness** of logical deduction systems, i.e. the 'provability' of all 'true theorems' in a given system of semantics [1, 2, 19, 20, 21, 22, 23]. The 'applied' motivation is the search for better understanding of the mathematical ideas underlying the Logic, Computer Science and Artificial Intelligence literature on automated theorem proving and reasoning, with special attention to the study of analogies and connections with Control theory.

In [8], I showed that some of the simpler textbook-style computer algorithms could be understood in terms of feedback control systems of a relatively simple algebraic structure. In this paper, I sketch a mathematical structure that is relevant to the study of properties of some of the logic-deduction systems to be found in the Computer Programming Theory, Knowledge Representation and Artificial Intelligence literature. These algorithms are inevitably more complicated in 'geometric' and 'algebraic' stucture than the algoritms considered in [8]. Since, in modern mathematics it has been found that one way to tame 'complication' is to introduce suitable algebraic structure, I will utilize the more algebraic aspects of Logic, with particular attention to the key property of 'completeness', following the Boolean-algebra based method used by Rasiowa and Sikorski [22,23].

In [10] (Chapter 1), I have outlined a general algebraic way of generalizing the basic ideas of the Ehresmann jet calculus - based theory of differential systems as from the manifold-theoretic framework where it originated to the discrete 'recursive function' world of the Logician and

Computer Scientist. In this paper, I use a more concrete algebraic version of the general definitions to 'geometrize' the formalism used in that part of Logic called **Proof Theory**. Mathematically speaking, we deal with families of maps of the integers into certain algebras whose quotients are Boolean algebras.

2. CONTROLLABILITY AND THE REACHABLE SET

Let us first formulate the 'reachability' concept in greater generality. Let T, X and Y be arbitrary sets. Set:

$$Z = \text{set of subsets of X.} \qquad (2.1)$$

Let $\Gamma(T, X)$ be the set of maps with domain T and range X. A **system of mappings** will be defined as a subset:

$$\Gamma \subset \Gamma(T, X) \qquad (2.2)$$

Definition. Let z, z' be elements of Z, i.e. subsets of X. z' will be said to be **reachable** from z relative to the mapping system Γ if there is a $\gamma \ \varepsilon \ \Gamma$ such that:

$$\gamma(z) = z' \qquad (2.3)$$

Let us specialize to Control Theory, using from the ordinary differential equations viewpoint [14] that is commonly used in the Control literature. Consider a control system of the following form:

$$dx/dt = f(x, u) \qquad (2.4)$$

$$x \ \varepsilon \ X = R^n, u \ \varepsilon \ U = R^m$$

x is the **state**, u the **control** or **input**. f is a smooth (i.e. infinitely differentiable) map

$$R^n \times R^m \ \text{-->} \ R^n \qquad (2.5)$$

Let $\Gamma(R, U)$ be the set of piecewise infinitely differentiable maps from the real numbers R to the manifold $U = R^m$. An element of $\Gamma(R, U)$ is then a curve $\mathbf{u}$: t --> u(t) ε U. Let $x_0 \varepsilon$ X and $\{t --> u(t)\} \varepsilon \Gamma(R, U)$. Solve the differential equation 2.1 for a curve in R^n :

$$t --> x(t) \qquad\qquad\qquad (2.6)$$

with initial condition:

$$x(0) = x_0. \qquad\qquad\qquad (2.7)$$

The curve 2.6 is called a **trajectory** of the control system 2.4. For each t ε R, the system 2.4 defines a map:

$$\Sigma(f, t): X \times \Gamma(R, U) --> X \qquad\qquad\qquad (2.8)$$

as follows:

$$\Sigma(f, t)(x_0, \mathbf{u}) = x(t), \qquad\qquad\qquad (2.9)$$

where t--> x(t) is the solution curve of control system indicated in 2.4.

Definition   The control system is **controllable at time t and initial position** x_0 if the following condition is satisfied:

$$\Sigma(f, t)[(\{x_0\} \times \Gamma(R, U)] = X \qquad\qquad\qquad (2.10)$$

Only when the the system 2.4 is linear are there reasonably definitive answers to the problem of finding sufficient conditions of an algorithmic nature for controllability. In the general, nonlinear, case, one can typically only find sufficient conditions of a 'local' nature, using a general theorem of differential system theory proved by W. L Chow [5] in 1939, and adapted to control-theoretic and geometric purposes by myself [7, 13], Jurdjevic and Sussmann [18, 24]. C. Lobry pointed out [19] the connection between the applicability of Chow's Theorem to the Controllability problem and the property of **reversibility** of control systems. We shall see that related questions also play a role in the study of Logic Deduction systems

The **extended controllability problem** is to say something of an algorithmic nature about the sets

$$\Sigma(f, t)[(\{x_0\} \times \Gamma(R, U)] \qquad (2.11)$$

called the **reachable sets**. Again, there is a result of differential system theory, the **singular foliation version of the Frobenius Complete Integrability Theorem** [7, 11], that is useful here to say something general about the geometric nature of these reachable sets in this continuous-time, smooth case. In my work [7, 12], in order to study this reachable set, I have focussed attention on the smooth real-valued function of the state space which are constant along the trajectories of the control system. They provide an a-priori limitation on the size of the reachable set, and the 'controllability' property is typically characterized by the algebraic conditions necessary to prove that the differential system for the functions constant along trajectories has no non-constant solutions. Let us review some of that material here.

Definition. A real-valued map:

$$\alpha: O \to R \qquad (2.12)$$

from an open subset O of the state space R^n of the control system 2.4 is said to be **conserved** relative to the system 2.4 if the following condition is satisfied:

The composite map $t \to \alpha(x(t))$ is locally constant. (2.13)

If a map α satisfying 2.12 is smooth, we can derive consequences from condition 2.13 by differentiating the map $t \to \alpha(x(t))$, and using the system equations 2.4 and condition 2.13:

$$0 = d/dt((x(t)) = \alpha_x dx/dt = \alpha_x f(x(t), u(t)). \qquad (2.14)$$

Set:

$$V^u = f(x, u)\, \partial/\partial x \qquad (2.15)$$

For each u, V^u is a vector field on R^n.

Theorem 2.1. A smooth map $\alpha: O \to R$ is conserved relative to the system 2.1 if and only if it satisfies the following system of first order linear partial differential equations:

$$V^u(\alpha) = 0 \text{ for all } u \, \varepsilon \, U. \qquad (2.16)$$

Proof. That α is conserved means, using 2.14, that:

$$\alpha_x(x(t))f(x(t), u(t)) = 0 \qquad (2.17)$$

for all input curves $t \to u(t)$. In particular, put in a constant input, obtaining:

$$\alpha_x(x)f(x, u) = 0 \text{ for all } (x, u) \, \varepsilon \, R^n \times R^m. \qquad (2.18)$$

Condition 2.18 is equivalent to 2.14.

Q.E.D.

Conditions 2.14 form a system of linear, first-order partial differential equations for the function $x \to \alpha(x)$. To see whether it has any solutions, it must be prolonged by differentating both sides of with respect to x. This has been described in [12], at least to the extent that an argument can be deduced using Chow's Theorem.

Let us now recall the jet-bundle formulation of path systems. Set:

$$\Gamma_{sm}(R, X) = \text{set of continuous,}$$
$$\text{piece-wise} \quad \text{infinitely differentiable maps,} \qquad (2.19)$$

$$\gamma : R \to X, \quad t \to \gamma(t)$$

For each positive integer r, introduce an equivalence relation on X x Γ(R, X) as follows:

(t, γ) and (t', γ') are equivalent if:

$$t = t' \qquad (2.20)$$

and

γ, γ' agree to the r-th order at t, $\qquad (2.21)$

i.e. in a local coordinate system for
X about γ (t) the derivatives of the curves
t --> γ (t) , t --> γ '(t) of order r or less agree at t.

The quotient of Γ_{sm}(R, X) by this equivalence relation is called the **space of r-th order jets**, denoted as:

$$J^r(R, X). \qquad (2.22)$$

It can be given a manifold structure.

Definition. An r-th order path system on X is defined by a subset

$$PS \subset J^r(R, X) \qquad (2.23)$$

An element $\gamma \in \Gamma$(R, X) is called a **trajectory** of the path system if the following condition is satisfied:

$$j^r(\gamma)(t) \in PS \quad \text{for all } t \in R, \qquad (2.24)$$

where $j^r(\gamma)(t)$ is the equivalence class of the equivalence relation 2.20-2.22 to which (t, γ) belongs. Let:

$$\Gamma(PS) = \text{the set of such trajectories of the path system PS} \qquad (2.25)$$

For the purposes of developing geometric analogies with ideas in Logic and Computer Science, it is useful to consider 'path systems' as **defined** by the collection $\Gamma(PS)$ of paths.

We can now formulate the notion of 'reachable set' in this setting:

Definition. Let PS $\subset J^r(R, X)$ be a path system, as defined above. Let X_0 be a subset of X. The **reachable set at time t of the path system PS, starting from X_0**, is the collection of points $\{\gamma(t)\}$, where γ runs over the set of trajectories of PS that lie on X_0 at time $t = 0$.

In case the subset PS of $J^r(R, X)$ is a submanifold with respect to the standard manifold structure on the jet spaces, the conditions 2.24 amount to a system of ordinary differential equations for the trajectory curves. In the classical literature on differential systems, such a system of first order is called a 'Monge System', which we shall now describe in terms of local coordinates.

Supose that:

$$X = R^n$$

$$T = R.$$

Let:

$$F: R^p \times R^p \longrightarrow R^q \tag{2.26}$$

be a smooth map. Consider the following system of ordinary differential equations:

$$F(z(t), dz/dt) = 0, \tag{2.27}$$

Such as system is called a **Monge System.** (In particular, feedback control systems are of this type.)

In [7, 12],   Chow's Theorem [5,] and a 'singular' version of the Frobenius Theorem for Completely Integrable Vector Field Systems [11] was used to derive a reachability result for the classical Monge systems. I will now reformulate in more coordinate-free terms an idea suggested in [7,12] as a way of estimating the reachable set.

Definition.   Let T and X be sets, and let Γ be a collection of maps: T --> X. Let ϕ: X --> Y be another map. ϕ is said to be **conserved** relative to the system Γ if the following condition   is satisfied:

For each $\gamma \, \varepsilon \, \Gamma$, the composite map

$$\phi\gamma : T \text{ --> } Y, \; t \text{ -->} \phi(\gamma \, (t)) \tag{2.28}$$

is constant as t ranges over T.

Theorem 2.2.   Let $\{R, X, PS\}$ be a path system, coordinatized by the real numbers R and taking values in the manifold X.   Let Φ be a collection of maps from X to Y,   such that each $\phi \, \varepsilon \, \Phi$ is conserved relative to the path system $\{R, X, PS\}$, in the sense of the above definition.   Let x_0 be a point of X.   Then, the reachable set at time t, starting from x_0,   of the path system $\{R, X, PS\}$ is contained in the following subset of X:

$$\{x \, \varepsilon \, X: \phi(x) = \phi(x_0) \text{ for all } \phi \, \varepsilon \, \Phi \, \} \tag{2.29}$$

As indicated in [7, 12], the Chow Theorem [5] will now give useful suffficient conditions for the set to be precisely the reachable set, in terms of the basic algebraic operation on vector fields called the Jacobi-Lie Bracket.

3. COMPLETENESS IN THE SENSE OF GODEL FOR LOGIC SYSTEMS

We can compare these 'controllability' problems with certain problems of Logic which center around the notion of **completeness** as it has arisen from Godel's famous work in the late 1920's and algebraically developed in the work of Rasiowa and Sikorski [22-24].

Definition. A **Logic** will be an ordered pair $(\mathbf{D}, \mathbf{L})$ consisting of a set $\mathbf{L}$ and a binary relation, denoted by $\mathbf{D}$, on $\mathbf{L}$. If

$$P \, \mathbf{D} \, Q, \tag{3.1}$$

for $P, Q \in \mathbf{L}$, we give to the relation 3.1 the meaning that "Q can be deduced from P". Given such a logic $(\mathbf{L}, \mathbf{D})$ a **sublogic** will be a subset $\mathbf{L}'$ of $\mathbf{L}$, together with the relation $\mathbf{D}$ restricted to $\mathbf{L}'$. For $\mathbf{L}' \subset \mathbf{L}$, set:

$$\mathbf{D}(\mathbf{L}', \mathbf{L}) = \{ P \in \mathbf{L} : \text{There exists a } P' \in \mathbf{L}' \text{ such that } P' \, \mathbf{D} \, P \} \tag{3.2}$$

$\mathbf{D}(\mathbf{L}', \mathbf{L})$ is called the **deductive closure of $\mathbf{L}'$ in $\mathbf{L}$**. Given the **logic $\mathbf{L}$**, consider a sequence:

$$\mathbf{L}_1, \mathbf{L}_2, \ldots, \mathbf{L}_m \tag{3.3}$$

of sublogics such that:

$$\mathbf{L}_{j+1} \subset \mathbf{D}(\mathbf{L}_1, \mathbf{L}_2, \ldots, \mathbf{L}_j, \mathbf{L}) \tag{3.4}$$
$$\text{for } 1 \leq j \leq m-1$$

We will then call $\mathbf{L}_1, \mathbf{L}_2, \ldots, \mathbf{L}_m, \ldots$ a **system of logics** [4]. The evolution of the individual propositions of the $\mathbf{L}_j$ will be called the **trajectories** of the system.

We may think of the trajectories of the sequence of sublogics 3.3 as the analogue of the family of trajectories $t \longrightarrow x(t)$ of the control system 2.1 . Finding the "deductive closure" is of the same 'geometric' nature as finding the "reachable set" associated with a control system .

One way thinking about this deductive closure is to introduce **semantics**. We will do this algebraically via the notion of 'valuation':

Definition. Let $(L, \mathbf{D})$ be a Logic. A **valuation** of this Logic is a mapping:

$$v: \mathbf{L} \longrightarrow \{\mathbf{T}, \mathbf{F}\} \qquad (3.5)$$

such that:

$$\text{If } P \ \mathbf{D} \ Q \text{ and } v(P) = \mathbf{T}, \text{ then } v(Q) = \mathbf{T} \qquad (3.6)$$

($\{\mathbf{T}, \mathbf{F}\}$ is the set with two elements, which for intuitive reasons we name "true" and "false".)

Godel's Completeness Theorem then says something about this 'reachable set' in terms of a set of valuations: In favorable cases, the elements of $\mathbf{L}$ that can be obtained via a trajectory of a logical system are those which take the value "$\mathbf{T}$" on certain sets of valuations. Note the similiarity in geometric structure to the characterization of the 'reachable set' of a smooth control systems by means of the smooth functions constant along trajectories.

Let us now turn to another algebraic version of the 'path system' idea.

4. DISCRETE PATH SYSTEMS AND SUBSETS OF FREE MONOIDS.

In Differential Geometry, the basic concepts of Path Geometry can be formulated in terms of the theory of Jet Spaces associated with manifolds. In this Section, I present an algebraic version of these geometric ideas, based on the algebraic structures called **free monoids.** I will indicate two ways of defining such path systems, by 'difference equations' and by 'fixed-point conditions.

Let us first recall what is traditionally meant by a 'path' in Differential Geometry.

Definition. A **path** in a manifold X is a map:

$$\gamma : [a, b] \longrightarrow X \qquad\qquad (4.1)$$
$$t \longrightarrow \gamma (t)$$

from an interval:

$$[a, b] = \{t: t \text{ is a real number and } a \leq t \leq b \}$$

of real numbers,   satisfying the following conditions:

$$\text{The map } \gamma \text{ is continuous} \qquad\qquad (4.2)$$

$$\gamma \text{ is piecewise infinitely differentiable.} \qquad\qquad (4.3)$$

If X is a real vector space,   the derivatives

$$t \longrightarrow d\gamma (t)/dt, \ d^2\gamma (t)/dt^2, \ldots \qquad\qquad (4.4)$$

can be defined in the usual way. A   **(smooth, continuous-time) path system** may be defined as the family of curves of form 4.1 satisfying a system of differential equations:

$$F(t,\ d\gamma\,(t)/dt,\ d^2\gamma\,(t)/dt^2,\ \ldots) = 0 \qquad\qquad (4.5)$$

The formulation of these concepts in the case where X is a more general geometric object than a vector space, e.g. a differentiable manifold, leads to the Ehresmann jet-bundle calculus.

Discrete versions of this concept may be formulated by discretizing the notion of 'time' in the usual way, replacing the real variable 't' with an integer variable 'n'. We then deal with families of maps:

$$n \text{-->} \gamma\,(n), \qquad\qquad (4.6)$$
$$a \leq n \leq b,$$

with systems of difference equations replacing the system 4.5 of differential equations.

However, let us proceed in a more algebraic way to study collections of discrete paths of type 4.6. We are led to consideration of ordered sequences $(\gamma\,(a),\ \gamma\,(a+1),\ \ldots\ )$. Such sequences may be identified with elements of the type of algebraic structure called a 'free monoid'; let us utilize this algebraic structure.

Definition. Let X be a set and let X* denote the free monoid constructed from X. An element γ of X* is a finite ordered sequence of elements of X, that we denote as:

$$\gamma = x_0 \ldots x_n \qquad (4.7)$$

Two such elements $\gamma = x_0 \ldots x_n$, $\gamma' = x_0' \ldots x_m'$ of X* are **equal** if and only if:

$$m = n, \text{ and } x_0 = x_0', \ldots, x_n = x_n' \qquad (4.8)$$

Their product is defined as:

$$\gamma \gamma' = x_0 \ldots x_n x_0' \ldots x_m' \qquad (4.9)$$

The identity element **e** of the monoid structure on X* is the empty set. A **discrete path system** on X is defined as a subset of X* . We usually denote path systems by the symbol 'Γ'.

Remark. The 'free monoids' are also the basic algebraic structure in Language Theory. Thus, what I have just called a 'discrete path system' could equally well be called a 'Language'. However, I believe that there is enough difference in 'pure' and 'applied' mathematical motivation to keep the parallel terminolgies distinct.

X* considered as an algebraic structure has a gradation:

$$\gamma \dashrightarrow |\gamma|$$

To define it , set:

$$|\gamma| = n+1, \qquad (4.10)$$

if γ is form 4.7. For each integer n, let:

$$X^*[n] = \{\gamma \,\varepsilon\, X^* : |\gamma| = n\}. \qquad (4.11)$$

If γ and γ' are as in 4.9, we then have:

$$|\gamma\gamma'| = n+m + 2 \qquad (4.12)$$

4.12 can be reformulated as the following relation:

$$X^*[n]X^*[m] \subset X^*[m+n] \qquad (4.13)$$

Remark. Note the 'geometric' intution that goes along with these definitions. The element $\gamma = x_0 \ldots x_n \,\varepsilon\, X^*$ is identified with a map of type 4.6:

$$\{0, 1, \ldots, n\} \dashrightarrow X, \qquad (4.14)$$
$$0 \dashrightarrow x_0, \; 1 \dashrightarrow x_1, \ldots, n \dashrightarrow x_n$$

The product 4.9 then coincides with the 'geometric' concatenation of paths.

Having identified elements of X^* with paths, we must define the algebraic analog of the **jet prolongations** in the geometric theory of smooth path systems. They will be maps from X^* to the Cartesian product of copies of X. We also want to develop a 'calculus' of such paths. In contemporary differential geometry, 'differential and integral calculus' is developed by identifying a collection of maps from a space on which the 'calculus' is to take place. We will follow the same idea. In our case, the maps are to be from subsets of X^* to Cartesian products of copies of X. Let us start off with the analogue of the 'coordinate functions' in the usual calculus of functions on R^n

Suppose that:

$$\gamma = x_0 \ldots x_n \qquad (4.15)$$

For each integer j, $0 \leq j \leq n$, let:

$$\delta(j)(\gamma) = x_j \qquad (4.16)$$

$\delta(j)$ is then defined by 4.16 as a map:

$$\delta(j): X^*[n+1] \longrightarrow X \qquad (4.17)$$

Let us now develop the algebraic analogue of the 'partial derivative' functions. For each integer k between 0 and n, and let $\Sigma(k, n)$ be the collection of one-one, order-preserving maps:

$$\sigma : \{1, ..., k\} \longrightarrow \{0, ..., n\} \qquad (4.18)$$
$$j \longrightarrow \sigma(j)$$

For each $\sigma \in \Sigma(k, n)$, define a map:

$$\delta(\sigma): X^*[n+1] \longrightarrow X^k \qquad (4.19)$$

by the following formula:

$$\delta(\sigma)(\gamma) = (x_{\sigma(1)}, ..., x_{\sigma(\kappa)}) \qquad (4.20)$$

We can now use these notations to define the discrete analog of smooth path systems, defined (e.g. as in 4.5) by systems of differential equations.

Definition. Let k and n be integers such that $0 < k < n$. For each $\sigma \in \Sigma(k, n)$, let $S(\sigma)$ be a subset of X^k. Set:

$$S = \{S(\sigma): \sigma \in \Sigma(k, n)\} \qquad (4.21)$$

$$\Gamma(S) = \{\gamma \in X^*[n] : \delta(\sigma)(\gamma) \in S(\sigma)\} \qquad (4.22)$$

Then, $\Gamma(S)$ is called the **discrete path system** whose equations are given by 4.22. Such a system of path equations is determined by the map

$$\sigma \longrightarrow S(\sigma).$$

Example. The path system defined by a discrete dynamical system.

Let

$$f: X \rightarrow X \qquad (4.22)$$

be a map. Associate with (4.22) the following first order difference equation:

$$x(i+1) = f(x(i)). \qquad (4.23)$$

An **orbit** of the system 4.23 is a map:

$$N \rightarrow X,$$
$$i \rightarrow x(i) ,$$

that satisfies 4.23. A **partially defined orbit** of the discrete dynamical system 4.23 is a map:

$$N \cap [a, b] \rightarrow X$$

with a, b real numbers, that satisfies 4.23 for all pairs (i, i+1) of integers such that:

$$(i, i+1) \, \varepsilon \, (N \cap [a, b]) \, x \, (N \cap [a, b]). \qquad (4.24)$$

Theorem 4.1. Let:

$$S = \{(x, x') \; \varepsilon \; X \; x \; X: f(x) = x'\} \tag{4.25}$$

and let:

$$\Gamma[n+1] = \{\gamma \; \varepsilon \; X^*[n+1]: (\delta(i)(\gamma) \; x \; \delta(i+1)(\gamma))[0, n] \subset S \text{ for } 0 \le i \le n-1\} \tag{4.26}$$

Then, the elements of $\Gamma[n+1]$ are the solutions $i \dashrightarrow x(i)$ of the difference equation 4.23 for the interval $0 \le i \le n$.

Proof. Follows from definitions.

In [10], I have indicated how the solutions of discrete-time dynamical systems of the form 4.23 may be defined by fixed-point conditions. (These fixed-point conditions may be considered as the discrete-equivalent of the classical 'Picard Fixed Point Equations' for solutions of continuous-time ordinary differential equations.) Here is another version of this formulation.

Theorem 4.2. Associate with the difference equation 4.23 the following map:

$$\phi: X^* \longrightarrow X^*.$$

$$\phi(x_0 \ldots x_n) = x_0\, f(x_0) \ldots f(x_{n-1}) \qquad (4.27)$$

Then, $\gamma = x_0 \ldots x_n \in X^*[n+1]$ satisfies the following following fixed-point equation:

$$\phi(\gamma) = \gamma \qquad (4.28)$$

if and only if the discrete path:

$$i \longrightarrow x_i$$

satisfies the difference equation 4.23.

Proof. 4.28 requires that:

$$\phi(x_0 \ldots x_n) = x_0 \ldots x_n \qquad (4.29)$$

Use 4.27 to obtain:

$$x_0\, f(x_0) \ldots f(x_{n-1}) = x_0 \ldots x_n \qquad (4.30)$$

The definition of 'equality' within X^* now requires that:

$$f(x_0) = x_1, \ldots, f(x_{n-1}) = x_n \qquad (4.31)$$

$$\textbf{Q.E.D.}$$

Here is another property of the map ϕ, defined by 4.27, and associated with the difference equation 4.23:

Theorem 4.3.   For each integer n, let:

$$\pi(n+1, n): X^*[n+1] \longrightarrow X^*[n] \qquad (4.32)$$

be the projection map defined by the following formula:

$$\pi(n+1, \ n)(x_0 \ldots x_n) = x_0 \ldots x_{n-1} \qquad (4.33)$$

Then, :

$$\phi\pi(n+1, n) = \pi(n+1, n)\phi. \qquad (4.34)$$

Proof. Using 4.33 and 4.27, we have:

$$\phi\pi(n+1, \ n)(x_0 \ldots x_n) = x_0 \ f(x_0) \ldots f(x_{n-2}) \qquad (4.35)$$

Also,

$$\pi(n+1, \ n)\phi(x_0 \ldots x_n) = \pi(n+1, \ n)(x_0 \ f(x_0) \ldots f(x_{n-1}))$$

$$=, \text{ using } 4.33 \text{ again,}$$

$$x_0 \ f(x_0) \ldots f(x_{n-2}) \qquad (4.36)$$

Equality of the right hand sids of 4.35 and 4.46 now finishes the proof of 4.34.
$$\textbf{Q.E.D.}$$

Remark. This result can be formulated in terms of Category Theory. Let **N** be the category associated with the ordered set consisting of the non-negative integers. Assign to each 'object' n of **N** the set $X^*[n+1]$, and to each 'morphism' (n, m), pairs of integers with $n \leq m$, the map:

$$\pi(m, n): X^*[m+1] \longrightarrow X^*[n+1]$$

$$\pi(m, n)(x_0 \ldots x_m) = x_0 \ldots x_n \tag{4.37}$$

This assigment defines a contravariant functor from the category **N** to the category **SET**. The mapping ϕ then defines a natural transformation between functors.

5. CONTROLLABILITY AND CONSERVED MAPS FOR DISCRETE PATH SYSTEMS

In this Section, we shall formulate the discrete-path version of the geometric ideas of 'controllability' and 'reachable set' associated with continuous-time systems and briefly discussed in Section 2.

Let X^* continue as the free monoid generated by the set X, as defined above.   We will consider path systems Γ which are subsets of X^*.

For $x_0 \, \epsilon \, X$, set:

$$X^*(x_0) = \{\gamma \, \epsilon \, X^* : \text{initial point of } \gamma = x_0\} \tag{5.1}$$

Definition. Let $\Gamma \subset X^*$ be a path system, as defined above. A point x is said to be **reachable** from x_0 relative to the path system Γ if there is a

$$\gamma \, \epsilon \, \Gamma \cap X^*(x_0) \tag{5.2}$$

such that:

$$\text{endpoint } (\gamma) = x. \tag{5.3}$$

Set:

$$R(x_0) = \{x \in X : x \text{ is reachable from } x_0 \text{ along a path in } \Gamma(x_0)\} \tag{5.4}$$

$R(x_0)$ is called the **set reachable from** x_0 associated with the path system Γ. Define a relation R on X, i.e. a subset of $X \times X$, as follows:

$$(x_0, x) \in R \quad \text{iff. } x \in R(x_0) \tag{5.5}$$

Theorem 5.1. The relation $R \subset X \times X$ defined by 5.5.is transitive (see Section 7 for the definition) if the following condition is satisfied:

$$\Gamma \text{ is a sub-monoid of } X^* \tag{5.6}$$

It is reflexive (again, see Section 7) if the path system Γ satisfies the condition:

> For each $x_0 \in X$, the discrete path with one element, x_0, belongs to Γ. $\tag{5.7}$

Proof. Let (x, x') and (x', x'') be points of $X \times X$ that lie in R. Proving transitivity of R requires constructing a path γ'' in Γ going from x to x''. Let γ be a path going from x to x', and γ' a path going from x' to x''. Set:

$$\gamma'' = \gamma \gamma' \tag{5.8}$$

Geometrically, γ'' is the concatenation of the paths γ' and γ. It then follows from 4.9 that γ'' goes from x to x'', thus implying reflexivity of R.

Q.E.D.

Definition. Let $\Gamma \subset X^*$ be a path system and let Y be a set. A map:

$$\alpha: X \to Y \tag{5.8}$$

is said to be a **conserved map** relative to the path system Γ if the following condition is satisfied:

For all $\gamma = x_0 \dots x_{n-1} \in \Gamma$

$$\alpha(x_0) = \alpha(x_i) \text{ for } 0 \le i \le n-1. \tag{5.9}$$

Geometrically, 5.9 means of course that the map $x \to \alpha(x)$ is constant along the discrete path in X associated with the element γ.

Theorem 5.2. Let $\Gamma \subset X^*$ be a path system, and let $\alpha: X \to Y$ be a map which is conserved relative to Γ. Let $x_0 \in X$. Then, the set reachable from x_0 along paths in Γ is contained in the set:

$$\alpha^{-1}(\alpha(x_0)) \tag{5.10}$$

Proof. Follows immediately from 5.9.

Remark. The set indicated in 5.10 may be denoted in geometric terminology as the **fiber of the map** α **above the image point** $\alpha(x_0)$.

6. SOME GENERAL SET-THEORETIC FACTS ABOUT RELATIONS AND QUOTIENT MAPS

In this Section, I will gather together some standard set-theoretic generalities about binary equivalence relations and associated maps. They play a basic role in the study of path systems.

Definition. Let X be a set. A **(binary) relation R** on X is a subset of X x X. The relation **R** is an **equivalence relation** if it is **reflexive** i.e. (x, x) ε **R** for x ε X; **transitive**, i.e. (x, x') ε **R** and (x', x'') ε **R** implies (x, x'') ε **R**; and **symmetric**, i.e. (x, x') ε **R** implies (x', x) ε **R**. The **symmetric part** of the relation **R** ⊂ X x X is the relation **R'** defined as follows:

$$\mathbf{R'} = \{(x', x): (x, x') \text{ and } (x', x) \text{ ε } \mathbf{R}\} \tag{6.1}$$

The **transitive closure of the relation R** ⊂ X x X is the relation **R"** defined as follows:

$$\mathbf{R"} = \{(x, x'): \text{The exists an integer n and an n-tuple} \tag{6.2}$$
$$(x_1, ..., x_n) \text{ of elements of X such that:}$$
$$(x, x_1), (x_1, x_2), ..., (x_n, x') \text{ ε } \mathbf{R}\}$$

Theorem 6.1. Let **R** ⊂ X x X be a relation on the set X and let **R"** be the transitive closure relation defined by formula 6.2. Then, **R"** is transitive.

Proof. Let (x, x') and (x', x'') ε **R"**. Then there exists integers n and m, an n-tuple $(x_1, ..., x_n)$, and an m-tuple $(y_1, ..., y_m)$, of elements of X such that:

$$(x, x_1), (x_1, x_2), ..., (x_n, x') \text{ ε } \mathbf{R} \tag{6.3}$$

$$(x', y_1), (y_1, y_2), ..., (y_m, x'') \text{ ε } \mathbf{R} \tag{6.4}$$

Putting these two chains together provides a chain:

$$(x, x_1), (x_1, x_2), \ldots, (x_n, x'), (x', y_1), (y_1, y_2), \ldots, (y_m, x'') \qquad (6.5)$$

of elements of **R** connecting x to x", completing the proof that **R**" is transitive.

Q.E.D.

Theorem 6.2. Let $\mathbf{R} \subset X \times X$ be a binary relation on the set X such that **R** contains the diagonal subset of $X \times X$. Let **R**' be the transitive closure of **R**, and let **R**" be the symmetric part of **R**'. Then, **R**" is an equivalence relation on X.

Proof. **R**" is reflexive, since it contains the diagonal subset of $X \times X$, and is symmetric, since it is the symmetric part of a relation. To complete the proof, we must show that **R**" is transitive.

Let (x, x') and $(x', x'') \varepsilon$ **R**". Since $\mathbf{R}'' \subset \mathbf{R}'$, and **R**' is transitive, we have:

$$(x, x'') \varepsilon \mathbf{R}'. \qquad (6.6)$$

Since **R**" is symmetric, we have:

$$(x', x) \text{ and } (x'', x') \varepsilon \mathbf{R}'' \subset \mathbf{R}' \qquad (6.7)$$

Transitivity of $\subset$ **R**' again implies that:

$$(x'', x) \varepsilon \mathbf{R}' \qquad (6.8)$$

6.6 and 6.8 now imply that:

$$(x, x'') \varepsilon \mathbf{R}'' \qquad (6.9)$$

which completes the proof of Theorem 6.2..

Definition. Let **R** be a relation on the set X. A map $\pi: X \to Y$ is said to be a **projection map relative to R** if the condition is satisfied:

$$\pi(x) = \pi(x') \text{ for all } (x, x') \in \mathbf{R} \qquad (6.9)$$

Theorem 6.3. Let **R** be a reflexive relation on the set X. Suppose that $\pi: X \to Y$ is a map which is a projection map relative to **R**. Let **R"** be the symmetric part of the transitive closure **R'** of **R**. Then π is a projection map relative to **R'** and **R"**.

Proof. Let us suppose that π satisfies 6.6. Let $(x_1, ..., x_n)$ be an n-tuple of elements of X such that:

$$(x, x_1), (x_1, x_2), ..., (x_n, x') \in \mathbf{R} \qquad (6.10)$$

Applying 6.9 to each term of 6.10, we have:

$$\pi(x) = \pi(x_1) = \pi(x_2) = ... = = \pi(x') \qquad (6.11)$$

This - together with the obvious fact that a projection map relative to a given relation is also a projection map relative to its symmetric part - implies that π is a projection map relative to **R'** and **R"**.

Q.E.D.

Definition. Let **R** be an equivalence relation on a set X. For x ε X, the **equivalence class of R containing x** is the subset:

$$\{x': (x, x') \, \varepsilon \, \mathbf{R} \} \tag{6.12}$$

An **equivalence class of R** is a subset of X which is the equivalence class of **R** of at least one element of X. Each point of X belongs to precisely one equivalence class. The **quotient of X by R**, denoted as:

$$X/\mathbf{R}, \tag{6.13}$$

is the set of equivalence classes of **R**. The map:

$$X \dashrightarrow X/\mathbf{R} \tag{6.14}$$

which assigns to each x ε X the equivalence class to which it belongs is called the **canonical projection map associated with the equivalence relation R**.

Remark. We will be particularly interested in reflexive relations $\mathbf{R} \subset X \times X$ such that the quotient:

$$X/ \text{ (symmetric part of the transitive closure of } \mathbf{R}) \tag{6.15}$$

carries certain algebraic or geometric structures. For example, for the equivalence relations generated by 'reachablity' for smooth path systems, the quotient 6.15 often carries a smooth manifold structure. In Propositional and Predicate Logic, where the relation **R** means, intuitively, that one element of X can be deduced from another by a rule of inference, the quotient 6.15 often carries a structure of a Boolean or Heyting algebra. Indeed, the existence of such an algebraic structure on the quotient 6.15 is the heart of the Rasiowa - Sikorski proof [22-24] of Godel Completeness for such logical systems. We now turn to study certain aspects of the algebra that occur in this work.

7. PROPOSITION ALGEBRAS AND THEIR VALUATIONS

The theory of the Propositional Calculus is a standard topic in every treatise on Logic [1, 20, 21]. Rather than defining it via recursions, i.e. 'well-formed formulas', I prefer to deal with algebraic structures which possess the desired properties: This is the approach of Algebraic Logic [6, 23, 24].

Definition. A **Proposition Algebra** is defined by an ordered triple:

$$\{X, \cup, \urcorner\} \tag{7.1}$$

consisiting of the following data:

$$X \text{ is a set} \tag{7.2}$$

$$\cup: X \times X \dashrightarrow X \text{ is a binary operation} \tag{7.3}$$

$$\urcorner: X \dashrightarrow X \text{ is a unary operation} \tag{7.4}$$

Define additional binary operations:

$$\cap: X \times X \dashrightarrow X \tag{7.5}$$

$$\Rightarrow: X \times X \dashrightarrow X \tag{7.6}$$

by the following formulas:

$$x \cap x' = \urcorner(\urcorner x \cup \urcorner x') \tag{7.7}$$

$$\Rightarrow(x, x') = (x \Rightarrow x') = \urcorner x \cup x' \tag{7.8}$$

The Boolean algebras [1, 2, 6, 15] form one class of Propositional Algebras. The algebras which occur in Propositional Logic [2, 20, 23, 24] are not Boolean, but are 'free' versions of them: They have quotients which are Boolean. (These quotients are the **Lindenbaum Algebras** [1, 2, 23]). They appear naturally when one seeks **valuations** of Propositional Algebras, i.e. homomorphisms into the **simple Boolean algebras**.

The 'axioms' for Boolean algebras are chosen to be algebraic properties of the 'model' Boolean algebra, namely the set of all subsets of a set, with the operation $\{\cup\}$ the union of subsets, $\{\,7\}$ the set-theoretic complement, and $\{\cap\}$ the intersection operation. Stone's Famous Theorem [1, 6, 15, 21] says that these are the only examples, i.e every Boolean algebra is isomorphic to a Boolean algebra of subsets of the set of **maximal filters** of the algebra. There is but one (up to isomorphism) Simple Boolean Algebra, namely the algebra of subsets of a set with one element. For intuitive purposes - and connections with Logic - we denote this algebra as follows:

$$\{\mathbf{T}, \mathbf{F}\} = \mathbf{BOOL} \tag{7.9}$$

and denote the two elements of this algebra as follows:

$$\mathbf{T} = \text{the whole set.}$$

$$\tag{7.10}$$

$$\mathbf{F} = \text{the empty set.}$$

A **valuation** of a Propositional Algebra $\{X, \cup, 7, \}$ is then defined by a mapping:

$$v: X \dashrightarrow \mathbf{BOOL} \tag{7.11}$$

which is a homomorphism between the algebraic structure defined by the operations $\cup, 7$ on X and the union-complementation operations on the set with one element.

8. TARSKI SEMANTICS.

Let X continue as a set which carries a family $V = \{v\}$ of maps:

$$v: X \dashrightarrow \{\mathbf{T}, \mathbf{F}\}. \tag{8.1}$$

Definition. For every subset Y of X, let:

$$V_Y = \{v \, \varepsilon \, V : v(y) = \mathbf{T} \text{ for all } y \, \varepsilon \, Y\} \tag{8.2}$$

Y is said to **semantically imply** $x \, \varepsilon \, X$ (with respect to V) if the following condition is satisfied:

$$v(x) = \mathbf{T} \text{ for all } v \, \varepsilon \, V_Y. \tag{8.3}$$

If 8.3 is satisfied for all $v \, \varepsilon \, V$ then x is said to be a **tautology** with respect to the valuations V . Set:

T(V) = set of tautologies with respect to the valuations V .
(8.4)

$$S_V(Y) = \{x \, \varepsilon \, X : Y \text{ semantically implies } x \text{ with } \tag{8.5}$$
$$\text{respect to the valuations } V\}$$

$S_V(Y)$ is called the **semantic closure** of the subset Y of X with respect to the set V of valuations.

The map:

$$Y \dashrightarrow S_V(Y) \tag{8.5}$$

is the basic mathematical object of the theory of Logical Deduction Systems. Think of it system-theoretically as a sort of 'input-output relation' . To a system-theorist [16], it is natural to want to construct a 'state-space realization', something like a 'dynamical system' which generates the map 8.5. Of course, to a computer scientist, it is important that such a 'dynamical system' have a 'computable' form. Here is a result which serves to determine such a 'realization' of the 'input-output' relation 8.5.

Theorem 8.1. Let $\{X, \cup, \urcorner\}$ be a Proposition Algebra, as derfined in Section 7. Let:

$$Y = \{y_1, ..., y_n\} \tag{8.6}$$

be a finite subset of X. Let $x \; \varepsilon \; X$. Set:

$$x_1 = \urcorner y_1 \cup x \tag{8.7}$$

$$x_2 = \urcorner y_2 \cup x_1 = \urcorner y_2 \cup (\urcorner y_1 \cup x) \tag{8.8}$$

$$\cdots$$

$$x_n = \urcorner y_n \cup x_{n-1} = \urcorner y_n \cup (\urcorner y_{n-1} \cup \; ... \tag{8.9}$$

Then:

$$x \; \varepsilon \; S_V(Y) \text{ iff } x_n \; \varepsilon \; T(V) \tag{8.10}$$

Also, the assignment:

$$x \; \text{-->} \; x_n$$

defines a map:

$$\beta: S_V(Y) \; \text{-->} \; T(V) \tag{8.11}$$

Proof. Suppose first that 8.6 is satisfied, i.e. that:

$$v(x) = \mathbf{T} \text{ for all } v \; \varepsilon \; S_V(Y) \tag{8.12}$$

Then, for $v \; \varepsilon \; \mathbf{V}(Y)$,

$$v(x_1) = v(\urcorner y_1 \cup x) = v(\urcorner y_1) \cup v(x) =, \text{ using 8.12,}$$

$$v(\neg y_1) \cup T = T \tag{8.13}$$

$$v(x_2) = v(\neg y_2 \cup x_1) = v(\neg y_2) \cup v(x_1) =, \text{ using } 8.13,$$

$$v(\neg y_2) \cup T = T \tag{8.14}$$

Continuing by induction:

$$v(x_n) = v(\neg y_n \cup x_{n-1}) = v(y_n) \cup v(x_{n-1}) = T \tag{8.15}$$
for all v satisfying 8.12.

Suppose now that:

$$v \ \varepsilon \ V - S_V(Y) \tag{8.16}$$

Then, for some integer r, $1 \leq r \leq n$, we have:

$$v(y_r) = F, \tag{8.17}$$

hence:

$$v(\neg y_r) = T \tag{8.18}$$

Using 8.7-8.8., we have:

$$v(x_r) = T, \tag{8.19}$$

hence, continuing using 8.7-8.9,

$$v(x_{r+1}) = T = \ldots = v(x_n). \tag{8.20}$$

This proves:

$$x_n \text{ is a tautology relative to the valuations } V . \tag{8.21}$$

Conversely, suppose that is satisfied. To complete the proof, we must show that:

$$x \, \varepsilon \, S_V(Y) \qquad (8.22)$$

If 8.22 is not satisfied, then there is a:

$$v_0 \, \varepsilon \, S_V(Y) \text{ such that: } v_0(x) = \mathbf{F}. \qquad (8.23)$$

8.23 implies that:

$$v_0(\urcorner \, y_1) = \ldots = v_0(\urcorner \, y_n) = \mathbf{F} \qquad (8.24)$$

Apply 8.23 and 8.24 to 8.7-8.9. We see then that:

$$v_0(x_n) = \mathbf{F} \qquad (8.25)$$

contradicting 8.21.

Q.E.D.

9. VALUATIONS AND SEMANTICS FOR BOOLEAN ALGEBRAS.

For a general Propositional Algebra, the practical usefulness of the results proved in Section 8 is limited by the difficulty in determing the set of all valuations. However, for the special type of Propositional Algebra called a 'Boolean Algebra', Stone's Theorem jumps in to cover this point. Let us then consider a Boolean algebra structure

$$\{B, \ \cup, \ \cap, \ \overline{}, 0, 1\} \tag{9.1}$$

Let $V(B)$ be the set of all valuations of the Boolean structure (9.1). To each finite collection $\{b_1, ..., b_n\}$ of elements of B, assign a subset of $V(B)$ as follows:

$$V(b_1, ..., b_n) = \{v \ \varepsilon \ V(B): v(b_1) = ... = v(b_n) = \mathbf{T}\} \tag{9.2}$$

In particular, we can assign to each $b \ \varepsilon \ B$ the subset:

$$V(b) = \{v \ \varepsilon \ V(B): v(b) = \mathbf{T}\} \tag{9.3}$$

Theorem 9.1.   The mapping $b \dashrightarrow V(b)$ is an isomomorphism of the Boolean algebra (9.1) onto a Boolean algebra of subsets of $V(B)$.

Proof. This is the Stone Theorem [1, 2, 6, 15, 21]. If B is not a finite set, it requires the Axiom of Choice.

Theorem 9.2. If $b \in B$ is such that:

$$v(b) = \mathbf{T} \text{ for all } v \in V(B) \tag{9.4}$$

then b is the element "1" of the Boolean structure 9.1.

Proof. B can be considered as a Boolean algebra of subsets of the set V(B). The only element of B that contains all points of V(B), i.e. takes the value "T" on all valuations, is the whole set , which is identified with the element "1" of the Boolean structure.

Remark. In terms of Logic, this says that the element "1" is the only "tautology" of a Boolean structure with respect to the set of all valuations.. This is then a special case of the Godel Completeness Theorem, for it tells how to effectively compute the Semantic Closure of the element "1" of a Boolean algebra. Notice however that even this simple result requires the Axiom of Choice!

Theorem 9.3. Let B continue as a Boolean structure. Let $\{b_1, ..., b_n\}$ be a finite collection of elements of B. Suppose that b is an element of B such that the following condition is satisfied:

$$v(b) = \mathbf{T} \text{ for all } v \in V(b_1, ..., b_n) \tag{9.5}$$

Then,

$$\neg b_1 \cup \neg b_2 \cup ... \cup \neg b_n \cup b = 1 \tag{9.6}$$

and the semantic closure of the set $\{b_1, ..., b_n\}$ in B is given by the following formula:

$$S_{V(B)}(\{b_1, ..., b_n\}) = \{b \in B: \neg b_1 \cup \neg b_2 \cup ... \cup \neg b_n \cup b = 1\} \tag{9.7}$$

Proof. A special case of Theorem 8.1.

Remark. Formula 9.6 motivates the introduction of the Heyting algebra [15] operation $\Rightarrow$ into the Boolean structure. $\Rightarrow$ is a binary operation: $B \times B \longrightarrow B$. Theorem 9.3 gives a precise meaning to the intuition behind the statement that:

$$(a \Rightarrow b) \text{ is 'true' means that "a logically implies b".} \qquad (9.8)$$

We can also construct a Path System in B which 'computes' the Semantic Closure of a finite number $\{b_1, ..., b_n\}$ of elements of B.

Let us now turn to showing how this determination of valuations for Boolean algebras can be used to Completeness of certain types of Deductive Systems.

10. PATH COMPLETENESS FOR PROPOSITIONAL ALGEBRAS USING BOOLEAN ALGEBRA METHODS.

Let:

$$\{X, \cup, \bar{}\} \qquad (10.1)$$

be a Propositional Algebra structure, and let

$$\{B, \cup, \cap, \bar{}, 0, 1\} \qquad (10.2)$$

be a Boolean Algebra structure. Suppose that:

$$\phi: X \longrightarrow B \qquad (10.3)$$

is a map that is a homomorphism in the Propositional Algebra sense. Let:

$$V(B) = \text{set of valuations of the Boolean algebra structure 10.1} \qquad (10.4)$$

$$V = \phi^*(V(B)) \qquad (10.5)$$

Set:

$$T(V) = \{x \, \varepsilon \, X: v(x) = \mathbf{T} \text{ for all } v \, \varepsilon \, V\} \qquad (10.6)$$

Theorem 10.1.

$$\phi^{-1}(1) \subset T(V), \qquad (10.7)$$

Proof. Let:

$$x \, \varepsilon \, \phi^{-1}(1). \qquad (10.8)$$

Then, for $v \, \varepsilon \, V$, there is a $v' \, \varepsilon \, V(B)$ such that:

$$v = \phi^*(v') \qquad (10.9)$$

Hence,

$$v(x) = \phi^*(v')(x) = v'(\phi(x)) =, \text{ using } 10.8, \, v'(1) = \mathbf{T} \qquad (10.10)$$

10.10 proves 10.7.

Q.E.D.

The following results are readily proved:

Theorem 10.2 Suppose that $A = \{a_1, \, ..., \, a_m\}$ is a finite subset of the Propositional Algebra $\{X, \}$. Let:

$$V(A) = \{v \, \varepsilon \, V : v(a) = \mathbf{T} \text{ for all } a \, \varepsilon \, A\} \qquad (10.11)$$

Set:

$$X(V, A) = \{x \, \varepsilon \, X: v(x) = \mathbf{T} \text{ for all } v \, \varepsilon \, V\} \qquad (10.12)$$

Then:

$$x \, \varepsilon \, X(V, A) \text{ iff. } (\neg a_1) \cup \ldots \cup (\neg a_m) \cup x \, \varepsilon \, T(V) \qquad (10.13)$$

Proof. Follows from .

Theorem 10. 3. Let

$$\Gamma \subset X^* \qquad (10.14)$$

be a path system on X. For $x_0 \, \varepsilon \, X$, let:

> $R(x_0)$ = set of points of X reachable on $\qquad$ (10.15)
> paths in Γ beginning at x_0

Suppose again that $A = \{a_1, \ldots, a_m\}$ is a finite subset of X. Suppose that the following conditions are satisfied:

$$T(V) \subset R(x_0) \qquad (10.16)$$

> For each integer j, $1 \leq j \leq m$, each $x \, \varepsilon \, X$,

$$\text{if } \neg a_j \cup x \ (= a_j \Rightarrow x) \, \varepsilon \, R(x_0), \text{ then } x \, \varepsilon \, R(x_0) \qquad (10.17)$$

Then,

$$\text{The semantic closure of } A \subset R(x_0) \qquad (10.18)$$

11. Input-output systems based on Boolean algebras.

In the case of Deductive Systems, the inputs and outputs are often finite sets consisting of well-formed-formulas of the Predicate Calculus. To model this algebraically, let B be a Boolean algebra, wth elements denoted as P, Q, P_1, etc. Denote the Boolean operations as follows:

$$(P, Q) \text{ ---> } P \cap Q \tag{11.1}$$

$$(P, Q) \text{ ---> } P \cup Q \tag{11.2}$$

$$P \text{ ---> } \sim P \tag{11.3}$$

Let X be the space of finite subsets of B . Denote an element x of X as follows:

$$x = \{P_1, ..., P_n\} \tag{11.4}$$

We are to think of X as the 'State Space' for the 'Deduction Systems' we are trying to construct.

DEFINITION. A *(Boolean) valuation* of a subset B' of B is a morphism from B to the True-False two element Boolean algebra which respects the Boolean operations to the extent they are defined on B.

Consider one such any valuation val of P to be fixed. Extend it to elements of X in the following way:

$$val \ (P_1, ..., P_n) = (val \ (P_1), ..., val \ (P_n)). \tag{11.5}$$

REMARK. It is useful for the theory of Knowledge Representation to leave open the possibility of 'valuations' taking values in other lattices, to cover 'default', 'many-valued' and 'fuzzy' logics.

DEFINITION. Let a set of valuations be given, and let X' be a subset of X. A map :

$$D: X' \dashrightarrow X \tag{11.6}$$

is called a ***deduction mapping*** if the following condition is satisfied:

> For every element $x \in X'$ such that the propositions
> that comprise it evaluate to **True**, the propositions $\tag{11.7}$
> that make up
> $D(x)$ also evaluate to **True**.

The main example of such a Deduction Mapping is the Robinson Resolution Rule.

DEFINITION. Let P be a given element of B Let X_P be the subset of X consisting of the elements x' of the following form:

$$x' = \{P \cup Q, {\sim}P \cup Q_1, ..., Q_m\}, \tag{11.8}$$

where $Q, Q_1, ..., Q_m$ are arbitrary elements of P.
Set:

$$D_P(x') = (Q \cup Q_1, ..., Q_m). \tag{11.9}$$

D_P is called the ***Resolution Map.***

THEOREM 11.1. Let D_P be the map defined above. Then, D is a deduction map in the sense that it maps collections of **True** propositions into **True** propositions.

Proof. Start of with an x of the form with all its constituent propositions **True**. Suppose that $P \cup Q$ and ${\sim}P \cup Q_1$ are both true, but that $Q \cup Q_1$ is **not True**. Then, both Q and Q_1 are **not True**, hence $P \cup Q$ and ${\sim}P \cup Q_1$ cannot both be **True**, which gives the contradiction.

Q.E.D.

12. 'SPEC' OF A BOOLEAN ALGEBRA.

One of the prime features of modern algebraic mathematics is the emphasis on **duality** as a unifying methodology. When one encounters an algebraic structure, it is often useful construct its 'dual'. What one means by this is made precise in Category Theory: Indeed, this unifying generality is one of the main contributions that Category Theory has made to mathematics as a whole. Booloean Algebras are the algebraic structures underlying the theory of Deductive Systems. Duality for these algebras was described by Marshall Stone in classic work of the 1930's that set the stage for much mathematics in succeeding years.   Let us formulate this in the following form:

Let **B** be a Boolean algebra. A *valuation* is a map:

$$v : \mathbf{B} \dashrightarrow \{t, f\} \tag{12.1}$$

which is a homomorphism in the Booloean algebra sense from **B** to the simplest, 2-element Boolean algebra generated by 'true' and 'false' in the logic sense. (Alternately, the right-hand side of 12.1 is the Boolean algebra of subsets of the set with one element. "**f**" is the empty subset and "**t**" is the whole set.) Let:

$$\mathbf{SPEC(B)} = \text{the set of valuation maps.} \tag{12.2}$$

The philosopher would think of **SPEC(P)** as the ' **set of all possible world interpretations**' for **P.** In analogous functional analysis and algebraic settings **SPEC(P)** is thought of as a 'spectrum', which explains our choice of notation. Part of Stone's Theorem implies that **P** is isomorphic as a Boolean algebra to the algebra of substs of **SPEC(P).**

We can now choose

$$X = \mathbf{SPEC(P)} \tag{12.3}$$

as the **state** **space** of Control Systems.   Such systems stand in the same relation to the theory of 'Default Logic' and its many variants as the space of systems with **B** as state space does to **Deduction** **Systems**.

This Resolution and Default process thus gives us a collection of partial maps acting on X, hence gives the possibility of organizing the Theorem Proving literature in a more system-theoretic way.   In order to be useful, it must be extended to the **Predicate** **Calculus** - introducing an additional complication, the presence of **Quantifiers** - and combined with the **Unification** algorithms.   I leave these Remarks at this point, to be continued later.

BIBLIOGRAPHY

[1]. J. L. Bell and M. Machover, A Course in Mathematical Logic, North-Holland, 1977

[2]. J. L. Bell and A. B. Slomson, Models and Ultraproducts, North-Holland, 1969

[3]. E. Cartan, Les systemes differentielles exterieures et leurs applications geometriques, Hermann, Paris, 1946.

[4]. P. E. Caines, R. Greiner, and S. Wang, Classical and Logic-based Dynamic Observers for Finite Automata, preprint, Mc Gill Univ., 1989.

[5]. W. L. Chow, Uber Systeme von linearen partiellen Differentialgleichungen erster Ordnung, Math. Ann., 117 (1939), 98-105

[6]. P. Halmos, Algebraic Logic, Chelsea Pub. Co., N. Y.,1962

[7]. R. Hermann, On the accessibility problem in control theory, in "International Symposium on Nonlinear Differential Equations and Nonlinear Mechanics", J. Lasalle and S. Lefschetz., eds., Academic press, 1963

[8]. R. Hermann, A 'geometric' view of the dynamics of the trajectories of computer programs, to appear, Acta Applicanda Mathematica

[9]. R. Hermann, E. Cartan's geometric theory of partial differential equations, *Adv. Math.* 1, 265-317 (1965)

[10]. R. Hermann, A category theory-based generalized differential system formalism for use in computer science, numerical analysis and control theory,

[11]. R. Hermann, The differential geometry of foliations, part 2, *J. Math. Mech.* (Indiana Math J.), 11, 303- 316 (1962).

[12]. R. Hermann, *Differential Geometry and the Calculus of Variations*, Academic Press, New York (1968). 2nd Edn., Interdisciplinary Mathematics, vol. 17, Math Sci Press, Brookline, MA (1977).

[13]. R. Hermann and A. Krener, Nonlinear controllability and observability, *IEEE Trans. Aut. Contr.*, **AC-22**, 728-740 (1977).

[14]. A. Isidori, *Nonlinear Control Systems: An Introduction*, Springer-Verlag, Berlin, 1985.

[15]. P. T. Johnstone, Stone Spaces, Cambridge University Press, 1982

[16]. R. E. Kalman, P. L. Falb, and M. A. Arbib. Topics in Mathematical System Theory, Mc Graw-Hill, 1969.

[17]. F. W. Lawvere, An elementary theory of the category of sets, Proc. Nat. Acad. Sci. USA, 52, 1964, 1506-1511

[18]. V. Jurdjevic and H. J. Sussmann , Controllability of nonlinear systems, J. Diff. Eq., **12**, 1972, 95-116.

[19]. C. Lobry, Controlabilite des systemes nonlineaires, SIAM J. Control **8**, 1970, 573-605

[20]. E. Mendelson, Introduction to Mathematical Logic, 3d Edition, Wadsworth, Monterey, Califonia, 1987

[21]. J. D. Monk, Mathematical Logic, Springer-Verlag, 1976

[22]. H. Rasiowa and R. Sikorski, A proof of the completeness theorem of Godel, Fundamenta Mathematica, 37, 1951, 193-200

[23]. H. Rasiowa and R. Sikorski, The Metamathematics of Mathematics, Warsaw, 1963.

[24]. H. Rasiowa, An Algebraic Approach to Non-classical Logics, North-Holland, 1974.

[25]. H. J. Sussmann, Lie brackets, real analyticity and geometric control, in "Differential Geometric Control Theory", R. Brockett, R. Millman, H. Sussmann, Eds., Birkhauser, 1983

[26]. A. Tarski, Der Wahrheitsbegriff in den formalisierten Sprache, Studia Philos., 1, 261-405 (1936). English translation in A. Tarski, "Logic, Semantics, Metamathematics", Edited by J. Corcoran. Second Edition, Hackett, 1983.

CARTAN: A COMPUTER LANGUAGE FOR KNOWLEDGE REPRESENTATION AND SYMBOLIC COMPUTATION IN MECHANICS AND CONTROL

1.   Introduction.

The aim of this chapter is to sketch how to combine mathematical ideas from Differential Geometry with those originating in the Knowledge Representation branch of Computer Science to develop a symbolic computation language for computation in the mathematical environment of Mechanics and Control.

In the past thirty years, differential geometry and Lie theory has been successfully applied to many areas of science and engineering. This mathematical methodolgy often involves the translation of the physical and engineering situation into computations in the algebraic system called Differential Grassmann Algebra.   It has been difficult to perform these computations in the computational enviroment routinely available to the scientist and engineer. The most success [5-7] has come from using the existing symbolic computation systems, such as MACSYMA, MAPLE, MATHEMATICA and REDUCE.   However, using these systems is often frustrated by the 'intermediate expression build-up' phenomenon. It seems likely that these computational difficulties can be surmounted by development of a computational system that works at a 'cleaner' and 'higher' algebraic level.

This research program is partially motivated by the development of Computational Logic [1, 33, 34], a branch of the theory of Knowledge Representation. The pioneer researchers in Artificial Intelligence underestimated the computational difficulties of Computational Logic, which in turn motivated the development of whole new areas of computer hardware, software and theory. The history of the language PROLOG is especially relevant, since it is a system dedicated to symbolic and algebraic computations in Propositional and Predicate Logic, an Algebraic System which is very similiar to the systems of Differential Grassmann Algebras which occur in the geometric theory of mechanics and control.

This Research Announcement begins the describes a prototype system for symbolic computation in mechanics and control based on the algebraic nature of the mathematical problems. In Computer Science terms, this might lead to the development of a theory of Knowledge Representation for these scientific and engineering disciplines. One mathematical foundation is the Method of The Moving Frame developed by the mathematician Elie Cartan in the early part of this century. The 'applied' goal is to develop flexible and powerful 'high-level' methods of description and analysis of mechanical and control systems encountered over a spectrum of engineering disciplines.

In the past thirty years much of the theory of Mechanics and Control has been reformulated in terms of differential-geometric concepts [2, 4, 15, 12, 23, 25, 28, 29] while computations are still handled in a very 19-th century framework.   Thus, in the study of the mathematics of the mechanics and control of a Robot or air-and-space craft, one deals with differential systems on manifolds built up from products of the Lie group of rigid motions and various of its subgroups, spaces which often have no globally-defined coordinate systems.   However, to do computations in currently-available computer algebra systems requires introducing local coordinate systems which - like the classical Euler angles - have computational pathologies of their own.   Anyone who has tried to even write down the equations of motion of a mechanical system consisting of coupled rigid bodies will have some feeling for the problem of 'intermediate expression build-up'!

Recall how the computational environment for the Propositional and Predicate Calculus of Logic is constructed [1, 18].   Start off with a (countable) alphabet A consisting of the following objects:

$$P, P_1, \cdots , \cap, \cup, \neg, (,) \tag{2.1}$$

Let A^* be the strings in this alphabet. The Propositional Calculus is a Language based on the alphabet A i.e. a subset $P(A)$ of A^* defined by certain well-know recursion rules, i.e. the so-called **well-formed formulas** of Logic.   A good deal of the literature of both the 'basic' and 'applied' Prolog/AI/Knowledge Representation literature is concerned with translating formulas in this language into do-able and efficient code, verifying 'meaning', correctness, etc. Let us briefly describe how the 'well-formed formulas' are constructed recursively.

The elements of A are called the **atomic propositions**. $\cap$ and $\cup$ are called the **disjunction** and **conjunction** symbols. $\neg$ is the **negation** symbol. Let A^* be the strings in this alphabet.   The following recursion rules determine the elements of $P(A)$:

The elements of $P(A)$ are well-formed formulas $\tag{2.2}$

If F, F' are well-formed formulas, then so are:

$$(F) \cap (F'), \ (F) \cup (F') \text{ and } \neg(F) \qquad (2.3)$$

$\mathbf{P}(A)$ is now defined as the smallest subset of A^* satisfying 2.2 and 2.3. .

We aim to analogously construct a Language to express Knowledge and Computation in Mechanics and Control in terms of a new Language related to the calculus of differential forms. This idea is based mathematically on Grassmann Algebras [4, 15, 25, 29, 30], just as the Propositional and Predicate Calculus is based on Boolean Algebras. These Grassmann Algebras come with additional 'differential' operators, analogously to the 'modal' operators of Logic.

Start with an Alphabet $\mathbf{C}$:

$$\mathbf{C} = \{f, f_1, .., \theta, \theta_1, ..., d, +, \wedge, (,)\}. \qquad (2.4)$$

f is the symbol for 'functions', or '0-forms'; θ the symbol for 1-forms; 'd' the symbol for 'exterior derivative'; $\wedge$ the symbol for 'exterior multiplication'. C' stands for 'Elie Cartan', the French mathematician active in the first half of the 20th century who developed the mathematical language of the 'Exterior Differential Calculus' , and applied it brilliantly in both 'pure' and 'applied' mathematics. A suitable Language $\mathbf{L}(\mathbf{C})$ will be constructed, that will be called: **CARTAN**.

The language $\mathbf{P}(A\ )$ of Propositional Logic may be considered as a 'free' version of the algebraic structures called Boolean Algebras, which can - by a basic theorem proved by Marshall Stone- be realized as Boolean algebras of subsets. There is a mapping of $\mathbf{P}(A\ )$ into a Boolean algebra called the **Lindenbaum-Tarski algebra** , which - combined with Stone's Theorem - leads to what the logicians call a **Model of the Language** $\mathbf{P}(A\ )$.

Similiarly, the Models associated to the Language $\mathbf{L}(\mathbf{C})$ are obtained in the following way: Let X be a differentiable manifold. For each integer r, let $\mathbf{D}^r(X)$ be the set of smooth differential forms of degree r on X. Let $\{f, \theta, d, +, \wedge\}$ be the symbols which occur in 2.2. Then: 'f' is associated with an element of $\mathbf{D}^0(X)$, 'θ' with an element of $\mathbf{D}^1(X)$, 'd' with the exterior derivative operation: $\mathbf{D}^r(X) \to \mathbf{D}^{r+1}(X)$, '$\wedge$' with the exterior multiplication operation $\mathbf{D}^r(X) \times \mathbf{D}^s(X) \to \mathbf{D}^{r+s}(X)$, etc. Our job as 'Knowledge Engineers' is now to translate information about mechanical and control systems into information in terms of the algebraic structure $\{\mathbf{D}^r(X), d, \wedge\}$, and then in terms of the Language $\mathbf{L}(C)$. This involves a process of 'Inverse Semantics'. The final step will be to 'translate' formulas in this Language into formulas in a known Programming Language such as LISP or PROLOG.

We can express some of the basic ideas in terms of the algebraic structure $\{\mathbf{D}^r(X), d, \wedge\}$:The 'inverse semantics' then expresses information about $\{\mathbf{D}^r(X), d, \wedge\}$ in terms of the Language $\mathbf{L}(C)$, or its equivalents within LISP or one of its dialects or specialized sub-language like MACSYMA. The basic Data Structure of the theory is the set of sequences $(\theta_1, ..., \theta_m)$ of elements of $\mathbf{D}^1(X)$. In Mechanics,one encounters these sequences in the form:

$$(\alpha_1, ..., \alpha_n, \beta_1, ..., \beta_n) \tag{2.5}$$

with the 2-form:

$$\omega = \Sigma \alpha_i \wedge \beta^i \tag{2.6}$$

defining the 'symplectic structure' associated with the mechanical system. This is done in a general field-theoretic framework in [10]

For example, consider a mechanical system of n degrees of freedom from the point of view of Hamiltonian Mechanics, in traditional form in terms of **configuration coordinates** $q = \{q^i: 1 \leq i \leq n\}$ and **momentum coordinates** $p = \{p_i\}$. Let $H(p, q)$ be a real-valued function of these coordinates, called the **Hamiltonian** or **Total Energy**. Set:

$$H_i = \partial H / \partial q^i, \quad H^i = \partial H / \partial p_i, \tag{2.7}$$

Then, the differential equations for the trajectory curves of the system are:

$$dq^i/dt = H^i, \quad dp_i/dt = -H_i. \tag{2.8}$$

Introduce the following differential forms:

$$\theta = \Sigma \, p_i dq^i - H dt \tag{2.9}$$

$$\omega = d\theta = \Sigma \, dp_i \wedge dq^i - dH \wedge dt. \tag{2.10}$$

Using 2.8, we have:

$$dH \wedge dt = \Sigma(H^i dp_i + H_i dq^i) \wedge dt \qquad (2.11)$$

Combining 2.10 amd 2.11, we have:

$$\omega = \Sigma\, dp_i \wedge dq^i - (H^i dp_i + H_i dq^i) \wedge dt. \qquad (2.12)$$

Using the algebraic rules for the Grassmann algebra operation '$\wedge$', we have:

$$\omega = \Sigma(dp_i + H_i \wedge dt) \wedge (dq^i - H^i \wedge dt). \qquad (2.13)$$

Set:

$$\alpha_i = dp_i + H_i \wedge dt, \; \beta^i = (dq^i - H^i \wedge dt. \qquad (2.14)$$

Then,

$$\omega = \Sigma \alpha_i \wedge \beta^i \qquad (2.15)$$

Hence:

**Passing from the Hamiltonian (i.e. total energy) function
H to the trajectory equations can be effected algebraically by
the factorization of the 2-form ω as a 'scalar product' (with
respect to the Grassmann algebra) of the n-vectors
$\alpha = (\alpha_1, ..., \alpha_n)$, $\beta = (\beta_1, ..., \beta_n)$**

Generalizing this approach to mechanical systems involving interactions and
constraints is the mathematical core of the Program.

Equally important is that many Control problems can be described in
the same mathematical 'language'. Consider a continuous-time feedback
control system in the form often considered in the engineering literature:

$$dx/dt = f(x, u), \; x \; \varepsilon \; R^n, \; u \; \varepsilon \; R^m \qquad (2.16)$$

Let $X = R^{n+m+1}$. Write: $x = (x^1, ..., x^n)$, $f = (f^1, ..., f^n)$. Introduce the following
1-forms on X:

$$\theta^1 = dx^1 - f^1 dt, \cdots, \theta^n = dx^n - f^n dt \qquad (2.17)$$

Many of the properties of the control system 2.16 can be translated into properties of the collection $(\theta^1, \cdots, \theta^n)$ of 1-forms, which can be then computed in terms of the Language $L(C)$ and the 'Inverse Semantics'. For example, a key calculation is that of the **derived system**, defined as the collection of 1-forms θ such that: $\theta \wedge \theta^1 \wedge \ldots \wedge \theta^n = 0$. This calculation plays a role in the theory of control via feedback linearization for nonlinear control systems.

The mathematics of 'feedback linearization' and construction of 'control laws' can also be described in terms of this Grassmann algebra. Consider a nonlinear control system of the form 2.16. A common engineering method of design of control is to utilize a linear system of the form:

$$dz/dt = Az + Bv, \ z \ \varepsilon \ R^n, \ v \ \varepsilon \ R^m \qquad (2.18)$$

Consider the following vector-valued 1- form on $R^{2n + 2m + 1 + n \times n}$:

$$\mu = dz - (Az + Bv)dt - \Lambda(dx - f(x, u)dt \qquad (2.19)$$

(The coefficients of the n x n matrix Λ are also variables.) Finding a **feedback linearization** [3, 26, 27, 36, 31, 32] of the nonlinear control system 2.12 amounts for finding n-dimensional submanifolds of $R^{2n + 2m + 1 + n \times n}$ on which the forms μ are zero. Utilizing the algorithms introduced in [26, 27, 36] for doing this exactly or pertubatively involves - at the first stage - computing the 'derived systems' of the system of 1 - forms which are the components of μ. As we have seen above, this computation is naturally formulatable in terms of Grassmann algebra.

The **optimal control problem** can also be expressed in a similiar framework. In its traditional form, it involves extremizing a 'performance criterion' in the form of a definite integral:

$$\int L(x, u)dt , \qquad (2.20)$$

subject to the control constraints 2.12. The 'differential form' method [23] for dealing with this problem is to define the 1- form:

$$\gamma = L(x, u)dt + \lambda(dx - f(x, u)dt) \qquad (2.21)$$

and then choosing the control variables u as a function of the variables (x, λ) (λ is the 'costate') so that:

$$d\gamma = \Sigma \alpha_i \wedge \beta^i, \qquad (2.22)$$

where the 1-forms α_i, β^i are chosen appropriately. Again, this leads to very elegant methods for computation which should translate into very efficient computer code.

3. DIFFERENTIAL GRADED ALGEBRAS AND THEIR DERIVATIONS.

One of my main goals is to introduce algebraic methodology for studying recursive systems that parallel methodology already developed by differential geometers to study differential systems. Inspired by the work of Elie Cartan, this might involve something like the algebra of differential forms. I will then review some of the underlying algebraic methodology of the Cartan theory,

Definition. A set D is a a **differential graded algebra** (over the real numbers) if it has the following algebraic structures:

D is a vector space over the real numbers. $\hspace{2cm}$ (3.1)

D has a graded vector space stucture

$$\{D^n: n = 0, 1, \dots \} \hspace{2cm} (3.2)$$

D has an associative-algebra structure:

$$(\theta, \theta') \longrightarrow \theta \wedge \theta', \hspace{2cm} (3.3)$$

satisfying:

$$D^n \wedge D^m \subset D^{n+m} \hspace{2cm} (3.4)$$

For $n = m = 0$, this product defines a ring structure on D^0. Following custom ,this product is dedined as:

$$(f, f') \longrightarrow ff' \quad \text{for } f, f' \, \varepsilon \, D^0. \hspace{2cm} (3.5)$$

For $n = 0$, $m \, \varepsilon \, N$, this product defines D^m as a module over the ring D^0, with the product defined as:

$$(f, \theta) = f\theta \text{ for } f \in D^0, \theta \in D^m \tag{3.6}$$

c). **D** is a **differential algebra,** i.e. has an operation:

$$d: D \to D, \tag{3.7}$$

called **exterior derivative**, satisfying the following conditions:

$$d^2 = 0 \tag{3.8}$$

d is R- linear $\tag{3.9}$

d is a **graded-derivation** of $\{D^n: n = 0, 1, \dots \}$, i.e.

$$d(\theta \wedge \theta') = d\theta \wedge \theta' + (-1)^m \theta \wedge d\theta' \tag{3.10}$$

for $\theta \in D^m, \theta' \in D$

Definition. Let $D = \{D^n: n \in N, d, \wedge\}$ be a graded differential algebra, as indicated above. A **vector field** relative to **D** is a map:

$$V: D \to D$$

satisfying the following conditions:

V is R-linear.

$$V(D^n) \subset D^n \qquad \text{for } n \in N$$

$$V(\theta \wedge \theta') = V\theta \wedge \theta' + \theta \wedge V(\theta') \text{ for } \theta, \theta' \in D.$$

$$V(d\theta) = dV(\theta) \text{ for } \theta \in D$$

There is an R - linear map

$$V \lrcorner : D \to D$$

$$\theta \dashrightarrow V \lrcorner \theta$$

such that:

$$V(\theta) = d(V \lrcorner \theta) + V \lrcorner d\theta \quad \text{for all } \theta \; \varepsilon \; \mathbf{D}$$

$$V \lrcorner (\theta \wedge \theta') = (V \lrcorner \theta) \wedge \theta' + (-1)^n \theta \wedge (V \lrcorner \theta')$$
$$\text{for } \theta \; \varepsilon \; \mathbf{D}^n, \theta \; \varepsilon \; \mathbf{D}$$

The operation $\theta \dashrightarrow V \lrcorner \theta$ is called the **inner product** or **contraction** operation.

Theorem Let $\mathbf{D}$ be a differential graded algebra. Let $\mathbf{V(D)}$ be the collection of vector fields. Then:

$\mathbf{V(D)}$ is a Lie algebra over the real numbers, with the Lie algebra operation [,] given by operator-commutator:

$$[V, \; V'](\theta) = VV'(\theta) - V'V(\theta)$$
$$\text{for } \theta \; \varepsilon \; \mathbf{D}, \; V, \; V' \; \varepsilon \; \mathbf{V(D)}$$

$\mathbf{V(D)}$ is a module over the ring $\mathbf{D}^0$:

$$(f, \; V) \dashrightarrow fV$$
$$(fV)(\theta) = f(V(\theta)). \text{ for } f \; \varepsilon \; \mathbf{D}^0, \theta \; \varepsilon \; \mathbf{D}$$

Proof. Direct verification, left to the reader.

Canonical example: The smooth differential forms on a finite dimensional, infinitely differentiable, paracompact manifold. Vector fields and Lie derivative.

Let X be a finite dimensional, infinitely differentiable, paracompact manifold. Then, it is a standard exercise in 'calculus on manifolds' to define a differential graded algebra:

$$\mathbf{D}(X) = \{\mathbf{D}^m(X): m = 0, 1, \dots \},$$

with the elements of $\mathbf{D}^m(X)$ identified with the infinitely differentiable differential forms of degree m. In differential geometric terms, $\mathbf{D}^m(X)$ can identified with the smooth cross-sections of the vector bundle on X whose fibers are the exterior product of m copies of the cotangent vector spaces to the manifold X. The operations 'd' and '∧' required to provide $\mathbf{D}(X)$ with a differential graded algebra structure - as defined above - are given by:

$$d = \text{exterior derivative operation}$$

$$\wedge = \text{exterior multiplication operation}$$

These operations were first codified in the work of Elie Cartan.

This 'geometric' realization of a differential graded algebra also has a Lie algebra attached. Set:

$$\mathbf{V}(X) = \text{set of smooth vector fields on X.}$$

Elements of $\mathbf{V}(X)$ are defined, geometrically, as smooth cross-sections of the tangent vector bundle T(X) to X. Each $V \in \mathbf{V}(X)$ defines a map:

$$\mathbf{D}(X) \dashrightarrow \mathbf{D}(X),$$
$$\theta \dashrightarrow V(\theta)$$

called **Lie derivative.** $\mathbf{V}(X)$ has a real Lie algebra structure:

$$(V, V') \dashrightarrow [V, V'],$$

called the **Jacobi-Lie bracket.**

Bibliography .

[1].   M. Fitting, First-Order Logic and Automated Theorem Proving, Springer-Verlag, 1990

[2].   R. Hermann, The differential geometric structure of general mechanical systems from a Lagrangian point of view, J. Math. Phy., 23, 2077- 2089, 1982.

[3].   R. B. Gardner, The Method of Equivalence and its Applications, SIAM, Philadelphia, PA., 1989.

[4].   P. Libermann and C-M. Marle, Symplectic Geometry and Analytical Mechanics, Reidel, 1987.

[5].   R. Grossman, Symbolic Computation: Applications to Scientific Computing, SIAM, 1989.

[6].   O. Akhrif and G. L. Blankenship, Symbolic computations in differential geometry applied to nonlinear control systems, in [5].

[7].   N Sreenath and P. S. Krishnaprasad, Mutibody simulation in an object-oriented programming environment, in [5].

[8].   R. Hermann, On the accessibility problem in control theory, in "International Symposium on Nonlinear Differential Equations and Nonlinear Mechanics", J. Lasalle and S. Lefschetz., eds., Academic press, 1963

[9].   R. Hermann, A 'geometric' view of the dynamics of the trajectories of computer programs, Acta Applicanda Mathematica, 18, 145-182, 1990.

[10].   R. Hermann, Differential form methods in the theory of variational systems and Lagrangian field theories, Acta Applicanda Mathematica **12**,
35-78 (1988)

[11].   R. Hermann, E. Cartan's geometric theory of partial differential equations, *Adv. Math.* 1, 265-317 (1965)

[9]. R. Hermann, A 'geometric' view of the dynamics of the trajectories of computer programs, Acta Applicanda Mathematica, 18, 145-182, 1990.

[10].   R. Hermann, Differential form methods in the theory of variational systems and Lagrangian field theories, Acta Applicanda Mathematica **12**,
35-78 (1988)

[11]. R. Hermann, E. Cartan's geometric theory of partial differential equations, *Adv. Math.* **1**, 265-317 (1965)

[12]. R. Hermann, Recursive Systems and Geometric Computer Science, Chapter 2.

[13]. R. Hermann, The differential geometry of foliations, part 2, *J. Math. Mech.* (Indiana Math J.), **11**, 303- 316 (1962).

[15]. R. Hermann, *Differential Geometry and the Calculus of Variations*, Academic Press, New York (1968). 2nd Edn., Interdisciplinary Mathematics, vol. 17, Math Sci Press, Brookline, MA (1977).

[16]. R. Hermann and A. Krener, Nonlinear controllability and observability, *IEEE Trans. Aut. Contr.*, **AC-22**, 728-740 (1977).

[17]. A. Isidori, *Nonlinear Control Systems: An Introduction*, Springer-Verlag,   Berlin, 1985.

[18]. E. Mendelson, Introduction to Mathematical Logic, 3d Edition, Wadsworth, Monterey, Califonia, 1987

[19]. H. J. Sussmann, Lie brackets, real analyticity and geometric control, in "Differential Geometric Control Theory", R. Brockett, R. Millman, H. Sussmann, Eds., Birkhauser, 1983

[20]. A. Tarski, Der Wahrheitsbegriff in den formalisierten Sprache, Studia Philos., 1, 261-405 (1936). English translation in A. Tarski, "Logic, Semantics, Metamathematics", Edited by J. Corcoran. Second Edition, Hackett, 1983.

[21]. R. Hermann, E. Cartan's geometric theory of partial differential equations, *Adv. Math.* **1**, 265-317 (1965)

[22].   R. Hermann, Cartan connections and the equivalence problem, *Cont. Diff. Eq.* **3**, 199-248 (1964)

[23].   R. Hermann, Some geometric aspects of the Lagrange variational problem, *Ill. J. Math* **6**, 634-673, 1962.

[24].   R. Hermann, The theory of equivalence of Pfaffian systems and input systems under feedback, *Math. Systems Theory*, **15**, 343-356 (1982).

[25].   R. Hermann, *Geometry, Physics and Systems*, Dekker, New York, 1973.

[26].   R. Hermann, Invariants for feedback equivalence and Cauchy characteristic multifoliations of nonlinear control systems, *Acta App. Math.,* **11** (1988), 123-153.

[27].   R. Hermann, Nonlinear feedback control and systems of partial differential equations,   *Acta App. Math.,* **17** (1989), 41-94.

[27a]. R. Hermann, Perturbation theory for nonlinear feedback control systems and Spencer-Goldschmidt integrability of linear partial differential equations, *Acta App. Math.,* **18** (1990), 417-58

[28].   R. Abraham and J. E. Marsden, *Foundations of Mechanics*, 2nd Edn., Addison-Wesley, Reading, MA 1979

[29].   E. Cartan, *Lecons sur les Invariants Integraux*,   Hermann, Paris, 1922

[30].   E. Cartan, *Les systemes differentielles exterieures et leurs applications geometriques*, Hermann, Paris, 1946.

[31].   R. Gardner, Differential geometric methods interfacing control theory, in R. Brockett, R. Millman and H. Sussmann (eds)., *Differential Geometric Control Theory,* Birkhauser, Boston, pp. 117-180.

[32].   L. R. Hunt, G. Meyer, and R. Su, Design for multi-input nonlinear systems, in R. S. Millman and H. Sussmann (eds), *Differential Geometric Control Theory*, Birkhauser, Boston, 1983.

[32].   H. Abelson and G. J. Sussman, Structure and Interpretation of Computer Programs, MIT Press, 1985.

[33] W. Bibel and Ph. Jorrand, Fundamentals of Artificial Intelligence, Springer-Verlag, 1986.

[34]. R. S. Boyer and J. S. Moore, A Computational Logic, Academic Press, 1979

[35]. M. R. Genesereth and N. J. Nilsson, Logical Foundations of Artificial Intelligence, Morgan Kaufmann Publishers, Inc., Los Altos, 1987

[36]. R. Featherstone, Robot Dynamics Algorithms, Kluwer, 1987.

[37]. R. Hermann, Lie Groups for Physicists, W. A. Benjamin, 1965.

MORE ABOUT JETS, GENERALIZED RECURSION AND APPROXIMATION OF DIFFERENTIAL EQUATIONS.

1. INTRODUCTION

As computers become more powerful and plentiful, they are being used with greater frequency as tools in science and engineering. My aim here is to develop foundations for such use, especially to study the mathematical structure.

From the point of view of the scientist and engineer, the theory of differential equations is one of the most fundamental and useful domains of mathematics. However, the computer science world is discrete, and the powerful mathematical tools and insights traditionally associated with differential equation theory have been lacking. Instead, computer science rests on the foundations of mathematical logic. Recursive function theory is the subdiscipline of mathematical logic of most direct connection to computer program theory, and it plays at least a background role in the theory of computer programming languages and algorithms. From the point of view of this paper, it is the link connecting the traditional differential equations viewpoint of the scientist and engineer and the mathematical-logic orientation of the computer scientist. One of my goals here is to bring the theory of recursive functions, and some of its extensions, more into the foreground as an analogue of the role that differential equations play in the traditional computation theory of science and engineering, and perhaps to prepare the way for its greater utilization. I will be particularly concerned with the internal algebraic structure of recursive functions and its relation to the theory of feedback control systems and some of the basic algorithms of computer programming theory.

Recursive function theory originated in the work of Hilbert, Ackermann, Gödel, Church, Turing, and Kleene as a by-product of their interest in the foundations of mathematics. It has become important for 'applied' reasons, because computer programs represent the 'practical' realization of the theoretical possibilities delineated by recursive function theory. Analogously, one of the main foundations of pure and applied differential geometry is the theory of differential systems, as developed in the work of Lie, Darboux, Goursat, Cartan, Vessiot, Ehresmann, Spencer, Goldschmidt, . . . I believe that there are relations between these two disciplines, and that making such relations more precise will be important and useful both for computer science and the various disciplines of applied mathematics.

Another of my underlying ideas is to link the theory of primitive recursive functions and the theory of integrable differential systems. In the theory of recursive functions, one of the main goals is to describe which functions F: $N^n \to N^n$, N = non-negative integers, are primitive recursive. Similarly, a main problem in the theory of differential systems has been to delineate the differential systems which can be solved in terms of systems of ordinary differential equations, and, among those, which have the property of 'integrability' or 'solubility by quadratures'. A "modern" formulation - pioneered by Spencer - has been to ask: Which differential systems defined over the category of infinitely differentiable manifolds can be solved while staying in this category?

On the mathematical/geometrical side, in this paper I want to see develop a variant of the standard Ehresmann-Spencer-Goldschmidt theory of differential systems based on the theory of jet spaces. My aim is to prepare the way geometrically (and Lie-theoretically) for studying the relation between systems of partial differential equations and their discretizations. We have seen that some Lie-theoretic principles are useful in the process of discretization of systems of ordinary differential equations, and control systems in particular.   I will begin to extend these methods to systems of partial differential equations.

Let us briefly indicate the Lie-theoretic methodology to be used in the discretization process. Let X be a finite dimensional, C^∞, paracompact manifold. Let

$$V_1, \dots, V_n$$

be a collection of vector fields on X. Suppose that

$$t \rightarrow g_i(t): \; X \rightarrow X \quad , \quad 1 \leq i, j \leq n$$

are one-parameter Lie groups acting on X whose infinitesimal generators are the $V_1, \ldots, V_n$, i.e.,

$$V_i = \frac{\partial}{\partial t} \, g_i(t) \, \Big|_{t=0} \quad ,$$
$$i = 1, \ldots, n \quad .$$

Let Y be a finite dimensional, real vector space. If

$$\gamma: \; X \rightarrow Y$$

is a smooth map, let

$$V_i(\gamma)(x) = \frac{\partial}{\partial t} \, \gamma(g(t)\,x) \, \Big|_{t=0} \quad ,$$

be the Lie derivative operation. Then,

$$\gamma \rightarrow V_i(\gamma)$$

is a first order, linear, differential operator.

Suppose given a vector-valued function

$$F(x, y, y_i, y_{ij}, y_{ijk}, \ldots)$$

of the indicated variables. Associate with F the following differential operator:

$$\gamma \rightarrow D(\gamma)$$

$$D(\gamma)(x) = F\left(x, \gamma(x), V_i(\gamma)(x), (V_i V_j)(\gamma)(x), \dots\right)$$

We can construct discrete-time approximations to V_i in the following way:

Choose one-parameter groups

$$t \to g_i'(t)$$

to approximate the basic Lie derivative operation

$$V_i(\gamma)(x) \sim h_i^{-1}\left[\gamma\left(g_i'(h_i)\, x\right) - \gamma(x)\right] \quad .$$

The real numbers $h_1, \dots, h_n$ are the **step-sizes**. We then approximate D by the following difference operator

$$\Delta(\gamma)(x) = F\left(x, V_i'(\gamma)(x), V_i' V_j'(\gamma)(x), \dots\right) \quad ,$$

with

$$V_i'(\gamma)(x) = h_i^{-1}\left[\gamma\left(g_i'(h_i)\, x\right) - \gamma(x)\right] \quad .$$

We obtain some of the classical discretization schemes for numerical approximations of partial differential equations by taking $V_1, \dots, V_n$ to be the directional derivatives $\dfrac{\partial}{\partial x^1}, \dots, \dfrac{\partial}{\partial x^n}$ with respect to a coordinate system $(x^1, \dots, x^n)$ for X. However, the geometric freedom to choose $V_1, \dots, V_n$ more generally as a moving frame leads to new mathematical possibilities which might be useful for various applied situations. What is particularly interesting from both a pure and applied point of view is that the availability of such symbolic computation systems as MACSYMA and MATHEMATICA opens up the possibilities for the user to carry out the formidable algebraic computations in a more feasible way.

2. ALGEBRAIC OPERATIONS ON THE SET OF MAPS

In preparation for such a generalized recursive function theory, let us describe various algebraic operations on spaces of mappings. Although this material is well-known - especially in the category theory literature it will be useful to have an explicit treatment in the context of the 'naive' set theory that is the lingua franca of the mathematician..

Let S be a collection of sets. Define:

$$\Gamma(S) = \{(\gamma, X, Y): \ X, Y \in S, \quad \gamma \text{ is a map: } X \to Y\} \tag{2.1}$$

Define a partial binary map

$$\Gamma(S) \times \Gamma(S) \to \Gamma(S)$$

$$(\gamma', X', Y'), (\gamma, X, Y) \to (\gamma'', X'', Y'') \tag{2.2}$$

via composition of mappings.

$(\gamma', X', Y') (\gamma, X, Y)$ is defined if and only if the following condition is satisfied:

$$Y \subset X' \quad . \tag{2.3}$$

If 2.3 is satisfied, set:

$$X'' = X \tag{2.4}$$

$$Y'' = Y' \tag{2.5}$$

$$\gamma'' = \gamma' \gamma \quad , \tag{2.6}$$

the composition of γ followed by γ'.

We then have:

$$(\gamma', X', Y') \, (\gamma, X, Y) = (\gamma' \, \gamma, X, Y') \tag{2.7}$$
if $Y \subset X'$.

This partial binary map satisfies the following associative law:

$$(\gamma'', X'', Y'') \, [(\gamma', X', Y'), (\gamma, X, Y)]$$
$$= [(\gamma'', X'', Y''), (\gamma', X', Y')] \, (\gamma, X, Y) \tag{2.8}$$

whenever (X, X', X'', Y, Y', Y'') satisfy the inclusion constraints needed to have the individual terms defined.

Remark. This algebraic structure is that of a **category**. The objects are the elements of S, and the identities attached to the objects are the triples

$$\left\{ (1_X, X, X)\colon X \in S \right\} \quad , \tag{2.9}$$

where "1_X" is the identity map attached to S.

Cartesian products introduce another algebraic structure. Suppose that S satisfies the following condition:

$$\text{If } X, Y \in S, \text{ then } X \times Y \in S \quad , \tag{2.10}$$

where

$$X \times Y = \left\{ (x, y)\colon x \in X, y \in Y \right\} \tag{2.11}$$

is the Cartesian product of the sets X and Y.

Define a mapping

$$\Gamma\,(\mathbf{S}) \times \Gamma\,(\mathbf{S}) \to \Gamma\,(\mathbf{S})$$

by the following formula:

$$(\gamma,\, X,\, Y)\,(\gamma',\, X',\, Y') = (\gamma \times \gamma',\, X \times X',\, Y \times Y') \quad , \tag{2.12}$$

where:

$$(\gamma \times \gamma')\,(x,\, x') = (\gamma\,(x),\, \gamma'\,(x')) \quad . \tag{2.13}$$

3. THE PRIMITIVE RECURSION OPERATION

The basic algebraic idea of recursive function theory is to study a collection of maps with two properties:

a) It is closed under composition, i.e., it forms a **category**;

b) It is closed under the two operations of **primitive recursion** and **recursive minimization**.

One of my aims in this paper is to emphasize more the relations between the theory of recursive relations and the theory of differential equations and systems.

Let
$$\mathbf{N} = \{n\colon\ n = 0,\, 1,\, 2,\, \dots \}$$
$$.$$

and let:

$$\delta\colon \mathbf{N} \to \mathbf{N}$$

be defined by

$$\delta\,(n) = n + 1 \quad . \tag{3.1}$$

δ is called the **iteration operator**. The **recursion operator** is 'dual', defined by:

$$\rho(n) = n\text{-}1)).$$

Let **S** continue as a collection of sets. Let

$$X, Y \in \mathbf{S} \quad . \tag{3.2}$$

Suppose that f is a map

$$X \times N \times Y \to Y \quad .$$

Consider a difference equation of the following form:

$$\gamma\,(x,\, n + 1) = f\,(x,\, n,\, \gamma\,(x,\, n)) \tag{3.3}$$

to be solved for a map

$$\gamma\colon X \times N \to Y \quad .$$

Theorem 3.1. Given maps

$$\alpha\colon X \to Y \quad ,$$
$$f\colon X \times N \times Y \to Y \quad ,$$

there is a unique map

$$\gamma\colon X \times N \to Y$$
$$(x,\, n) \to \gamma\,(x,\, n)$$

which satisfies the difference equation (3.3), and the following initial condition:

$$\gamma\,(x,\, 0) = \alpha\,(x) \quad . \tag{3.4}$$

The assignment

$$(\alpha,\ f) \to \gamma \qquad\qquad (3.5)$$

then defines a map on function spaces, which is called the **primitive recursion operation**.

Proof. Obvious, by induction on n.

The **primitive recursive functions** are those built up iteratively from solutions of primitive recursion equations of form (3.3), starting with **S** as the collection of sets

$$\mathbf{Z}^n,$$

where $\mathbf{Z}$ = ring of integers.

Such solutions of difference equations are to be combined with composition maps already defined as primitive recursive.

Let us develop relations with the theory of differential systems.

4. THE JET SPACES ASSOCIATED WITH A VECTOR FIELD SYSTEM

The geometric theory of differential systems is based on Ehresmann's theory of jet spaces. It serves as a kinematic background for the analytic and algebraic theory needed to treat the differential equations that arise in geometric applications. I will briefly recall the basic definitions of the Ehresmann theory.

Let X and Y be finite dimensional, infinitely differentiable, paracompact manifolds. ("Paracompact" means that they can be exhibited as a union of a countable number of compact subsets.) Set:

$$\Gamma\ (X,\ Y) = \text{pre-sheaf of local } C^\infty \text{ maps } X \to Y\ . \qquad (4.1)$$

More concretely, an element of $\Gamma(X, Y)$ is an ordered triple:

$$(\gamma, O, x)$$

satisfying the following conditions:

    a) O is an open subset of X;   (4.2)

    b) $x \in O$;   (4.3)

    c) γ is a C^∞ map: $X' \to Y$.   (4.4)

For each positive integer r, introduce an equivalence relation $\sim$ in $\Gamma(X, Y)$ as follows:

$$(\gamma, O, x) \sim (\gamma', O', x')$$

if and only if the following conditions are satisfied:

    $x = x'$,   (4.5)

    γ and γ' agree to order r at x, i.e., in a local coordinate system about x, all partial derivatives of γ and γ' of order at most r agree at x .   (4.6)

Theorem 4.1. Relations (4.1) — (4.6) define an equivalence relation on $\Gamma(X, Y)$.

Proof. Follows routinely from the definitions. The details are left to the reader.

Definition. The quotient space of Γ (X, Y) under the equivalence relation (4.5) — (4.6) is denoted as follows:

$$J^r (X, Y) \tag{4.7}$$

and called the space of **r-jets of maps X $\to$ Y**.

We can also use Γ (X, Y) to define projection maps between the jet spaces.

Define maps

$$\pi_X: \Gamma (X, Y) \to X \tag{4.8}$$
$$\pi_Y: \Gamma (X, Y) \to Y \tag{4.9}$$

as follows:

$$\pi_X (\gamma, O, x) = x \quad , \tag{4.10}$$
$$\pi_Y (\gamma, O, x) = \gamma (x) \quad . \tag{4.11}$$

Theorem 4.2. π_X and π_Y are both constant on the equivalence classes of the relations (4.5) — (4.6), hence pass to the quotient to define maps

$$\pi_X: J^r (X, Y) \to X \tag{4.12}$$
$$\pi_Y: J^r (X, Y) \to Y \quad . \tag{4.13}$$

(4.12) is called the **source**; (4.13) the **target mapping**.

Proof. That π_X and π_Y pass to the quotient is evident from (4.10) — (4.11).

In the Spencer–Goldschmidt theory of differential systems, the algebraic structure of the jet-spaces plays a dominant role. In local coordinate systems, the fibers of the source fiber space (4.12) can be identified with iterated symmetric tensor products. Classically, they are identical to the polynomials in the Taylor expansions used to identify the "order of contact" of elements of $\Gamma (X, Y)$. In the next section, I describe a coordinate-free approach to the description of this algebraic structure.

5. THE ALGEBRAIC STRUCTURE OF THE JET-SPACES FOR MAPS BETWEEN VECTOR SPACES

Specialize the manifolds X and Y to be finite dimensional, real vector spaces. For $x \in X$, let

$$V(x)$$

be the vector field on X which is the infinitesimal generator of the one-parameter group of translations generated by:

$$T_t (x') = x' + tx \quad . \tag{5.1}$$

If

$$\gamma\colon O \to Y$$

is a C^∞ map from an open subset O of X to Y, then

$$V (x) (\gamma) (x') = \frac{\partial}{\partial t} \gamma (x' + tx) \bigg|_{t = 0} \quad , \tag{5.2}$$

for $x' \in O$.

We recognize γ_r as the algebraic form of the **directional** **derivative** of γ in direction x.

Theorem 5.1. The map

$$x \to V (x)$$

is a one-one, linear map of X into the Lie algebra of vector fields on X. It satisfies the following commutativity condition:

$$V(x_1) \, V(x_2) = V(x_2) \, V(x_1) \tag{5.3}$$

for $x_1, x_2 \in X$.

In Lie algebraic terms,

$$x \to V(x)$$

is a one-one homomorphism from X (made into an abelian Lie algebra) into the Lie algebra of vector fields on the manifold X.

Proof. From $x_1, x_2, x' \in X$, using (5.2):

$$
\begin{aligned}
V(x_1)\, &V(x_2)\, (\gamma)\, (x') \\
&= \frac{\partial}{\partial t_1} \left(V(x_2)\, (\gamma)\, (x' + t_1\, x_1) \right) \Big|_{t_1 = 0} \quad , \\
&= \frac{\partial}{\partial t_1} \left(\frac{\partial}{\partial t_2} \left(\gamma(x' + t_1\, x_1 + t_2\, x_2) \Big|_{t_2 = 0} \right) \right) \Big|_{t_1 = 0} \quad , \\
&= \frac{\partial}{\partial t_1\, \partial t_2} \left(\gamma(x' + t_1\, x_1 + t_2\, x_2) \right) \Big|_{t = 0} \quad , \tag{5.4} \\
&= \frac{\partial^2}{\partial t^2} \left(\gamma(x' + t(x_1 + x_2)) \right) \Big|_{t = 0} \quad , \tag{5.5} \\
&= V(x_2)\, V(x_1)\, (\gamma)\, (x') \quad .
\end{aligned}
$$

This proves (5.3).

Q.E.D.

Theorem 5.2. Let

$$\gamma, \gamma': O \to Y$$

be C^∞ maps from an open subset O of X, and let

$$x \in O \quad .$$

Then, γ and γ' agree to the r-th order at x if and only if the following conditions are satisfied:

$$\gamma\,(x) = \gamma'\,(x) \quad . \tag{5.5}$$

For each integer s,
$$0 < s \le r \quad , \tag{5.6}$$
and each s-tuple
$$(x_1, \dots , x_s)$$
of elements of X, we have:

$$V\,(x_1) \dots V\,(x_s)\,(\gamma)\,(x) = V\,(x_1) \dots V\,(x_s)\,(\gamma')\,(x). \tag{5.7}$$

Proof. Expanding formula (5.5) by induction on s, we have
$$\frac{\partial^s}{\partial t^s}\left(\gamma\,(x + t\,x_1 + \dots + t\,x_s)\right)\bigg|_{t=0} \quad ,$$
$$= V\,(x_1) \dots V\,(x_s)\,(\gamma)\,(x) \quad . \tag{5.8}$$

Using (5.7), we see that, geometrically, conditions (5.5) and (5.6) say that the maps γ and γ' agree to the r-th order on each line passing through x. Using calculus (i.e., Taylor's formula), we see that this implies that γ and γ' agree to r-th order at x in the manifold sense, i.e., that in each local coordinate system about x, the partial derivatives of order r agree.

Q.E.D.

We can now use this identification of the vector space X with an abelian Lie algebra of vector fields on X to construct local coordinates for $J^r\,(X)$.

Let $x \in O \subset X$, and let

$$\gamma\colon O \to Y$$

be a C^∞ map. Then,

$$(\gamma, O, x) \in \Gamma (X, Y) \quad . \tag{5.9}$$

For each integer s satisfying (5.6), each s-tuple

$$(x_1, \dots , x_s) \in X \times \dots \times X \quad ,$$

assign to (γ, O, x) the element

$$V (x_1) \dots V (x_s) (\gamma) (x) \in Y \quad .$$

For fixed $(\gamma, O, x) \in \Gamma (X, Y)$, this defines a map:

$$\alpha (s, \gamma, O, x): \ X \times \dots \times X \to Y \tag{5.10}$$

$$(x_1, \dots , x_s) \to V (x_1) \dots V (x_s) (\gamma) (x) \quad . \tag{5.11}$$

Theorem 5.3. The map $\alpha (s, \gamma, O, x)$ defined by (5.11) is a real-multilinear map of the vector space X to the vector space Y. It depends **symmetrically** on $x_1, \dots , x_s$, hence defines an element of

$$S^s (X^d) \otimes Y \quad , \tag{5.12}$$

where X^d denotes the dual vector space to X and $S^s (X^d)$ is the s-fold symmetric tensor product of s copies of the vector space X^d.

Proof. Follows from (5.10) — (5.11) and the commutativity of the vector fields $\{ V (x): x \in X \}$. The identification of the s-fold, multilinear, symmetric maps

$$X \times \dots \times X \to Y$$

with the vector space (5.12) is standard multilinear algebra.

Theorem 5.4. For each integer $s \leq r$, the map

$$(\gamma, O, x) \to \alpha\,(s, \gamma, O, x) \in S^s\,(X^d) \otimes Y \tag{5.13}$$

passes to the quotient to define a map

$$\alpha\,(s, r)\colon J^r\,(X, Y) \to S^s\,(X^d) \otimes Y \quad. \tag{5.14}$$

Proof. Follows from Theorem 5.2.

We can now elucidate the algebraic structure of the fiber space

$$\pi_X\colon J^r\,(X, Y) \to X \quad.$$

Theorem 5.5. Let $Z\,(s)$ be the real vector space which is the direct sum of the vector spaces

$$S^s\,(X^d) \otimes Y \quad, \tag{5.15}$$
$$0 \leq s \leq r \quad.$$

Let:

$$E\,(r) = X \times Z\,(r) \tag{5.16}$$

be the product fiber space. Define a map

$$\alpha\,(r)\colon \Gamma\,(X, Y) \to E\,(r) \tag{5.17}$$

by the following formula:

$$\begin{aligned}
&\alpha\,(r)\,(\gamma, O, x) \\
&= (x, \gamma\,(x), \alpha\,(1)\,(\gamma, O, x), \alpha\,(2)\,(\gamma, O, x), \ldots, \alpha\,(r)\,(\gamma, O, x)) \quad.
\end{aligned} \tag{5.18}$$

Then, $\alpha\,(r)$ passes to the quotient to define a map

$$\alpha \, (r): \; J^r \, (X, \, Y) \to E \, (r) \qquad . \tag{5.19}$$

$\alpha \, (r)$ is an isomorphism between fiber spaces.

Proof. That $\alpha \, (r)$ is constant on the equivalence classes of $\Gamma \, (X, \, Y)$ whose quotient is $J^r \, (X, \, Y)$ should be clear from what has gone before. The proof that $\alpha \, (r)$ is a fiber space isomorphism follows from the Taylor expansion of the map γ in $(\gamma, \, O, \, x)$ about the point x.

$$\textbf{Q.E.D.}$$

6. DIFFERENTIAL SYSTEMS BASED ON THE THEORY OF JET SPACES

Our main interest in the theory of jet spaces is that it is an essential ingredient in any systematic geometric theory of differential systems. (The Ehresmann theory also plays a key role in other branches of differential geometry, such as the theory of mapping singularities.)

Let X and Y continue as manifolds, as defined in previous sections. Let

$$J^r \, (X, \, Y)$$

be the space of r-jets of smooth maps from X to Y.

Let

$$\gamma : O \to Y$$

be a C^∞ map of the open subset O of X. For each $x \in O$, consider the element

$$(\gamma, O, x) \in \Gamma \, (X, \, Y) \qquad . \tag{6.1}$$

Assign to (6.1) the equivalence class on $J^r \, (X, \, Y)$ to which it belongs. We obtain an element of $J^r \, (X, \, Y)$ which we denote as:

$$j^r (\gamma) (x) \quad . \tag{6.2}$$

Theorem 6.1. As x varies over O, the assignment

$$x \to j^r (\gamma) (x) \tag{6.3}$$

defines a mapping

$$j^r (\gamma): \; O \to J^r (X, Y) \tag{6.4}$$

such that:

$$\pi_X \; j^r (\gamma) = \text{identity} \quad \text{map} \tag{6.5}$$

$$\pi_Y \; j^r (\gamma) = \gamma \quad . \tag{6.6}$$

Proof. Follows from the definitions.

Definition. An r-th order differential system, with domain X, range Y, is defined by a submanifold

$$\mathbf{DS} \subset J^r (X, Y) \quad . \tag{6.7}$$

A smooth map

$$\gamma: \; O \to Y$$

from an open subset $\mathbf{X}$ of X to Y is said to be a **solution** of the system (6.7) if the following condition is satisfied:

$$j^r (\gamma) (O) \subset \mathbf{DS} \quad . \tag{6.8}$$

7.   DIFFERENTIAL OPERATORS

Having given the concept of "differential system" a precise geometric (and categorical) meaning, let us consider the related classical concept, that of differential operators.

Definition. Let X, Y, Z be manifolds. An **r-th order differential operator** taking Y-valued functions to Z-valued functions of X is a smooth map:

$$D: J^r(X, Y) \to Z \quad . \tag{7.1}$$

Let us show how this definition leads to the more classical notion of a mapping from functions to functions.

Let

$$\gamma: O \to Y$$

be a smooth map from an open subset O of X to Y. Its r-jet $j^r(\gamma)$ is a map

$$j^r(\gamma): O \to J^r(X, Y) \quad . \tag{7.2}$$

Define:

$$\mathbf{D}(\gamma) = D \; j^r(\gamma) \quad . \tag{7.3}$$

The map

$$\gamma \to \mathbf{D}(\gamma)$$

is then the operator on functions canonically associated with the map indicated by (7.1).

Here is a natural relation between the "differential equation" and "differential system" concepts.

Theorem 7.1. Let $y_0 \in Y$, and

$$\mathbf{DS}\,(D,\,y_0) = D^{-1}\,(y_0) \subset J^r\,(X,\,Y) \quad . \tag{7.4}$$

Then, a map $\gamma\colon O \to Y$ is a solution of the differential system (7.4) if and only if it satisfies the following **differential equation**:

$$\mathbf{D}\,(\gamma)\,(x) = y_0 \tag{7.5}$$
for all $x \in O$.

Proof. Follows again from definitions.

We now turn to finding an analogous geometric meaning to "discretization."

8. EULER DISCRETIZATION OF DIFFERENTIAL OPERATORS ON VECTOR SPACES

Return to the special case that X and Y are finite dimensional, real vector spaces. The algebraic foundation of our description of the jet spaces

$$J^r\,(X,\,Y)$$

was the map

$$x \to V\,(x) \tag{8.1}$$

from X to the vector space of first-order linear differential operators on the pre-sheaf $\Gamma\,(X,\,Y)$.

Let us "discretize" $V\,(x)$, using the simplest Eulerian method. Recall that

$$V (x) (\gamma) (x') = \lim_{h \to 0} [\gamma (x' + hx) - \gamma (x')] \, h^{-1} \quad .$$

For each $x \in X$, set:

$$\delta (x) (\gamma) (x') = \gamma (x' + x) \tag{8.2}$$

$$\delta (x): \Gamma (X, Y) \to \Gamma (X, Y) \quad . \tag{8.3}$$

Then

$$V (x) (\gamma) = \lim_{h \to 0} h^{-1} (\delta (hx) (\gamma) - \gamma) \quad . \tag{8.4}$$

Hence, (without being precise about which topology on operators we are using)

$$V (x) = \lim_{h \to 0} h^{-1} [\delta (hx) - 1] \quad , \tag{8.5}$$

where **1** is the identity operator.

Let:

$$M (X, Y) = \text{vector space of maps:} \quad X \to Y \quad . \tag{8.6}$$

Definition. For each real number h, let $V (x, h)$ be the linear map

$$M(X, Y) \to M (X, Y)$$

defined by the following formula:

$$V (x, h) (\gamma) = h^{-1} (\delta (hx) (\gamma) - \gamma) \quad . \tag{8.7}$$

$V (x, h)$ is called the **Euler discretization** of the first-order differential operator $V (x)$.

We can extend this discretization procedure to higher order operators which result from products

$$V(x_1) \ldots V(x_r)$$

of the basic vector fields.

Let us now turn as an example to the discretization of first-order partial differential equations.

9. DISCRETIZATION OF SYSTEMS OF FIRST-ORDER PARTIAL DIFFERENTIAL EQUATIONS

Consider a partial differential equation of the following form:

$$F\left(\frac{\partial \gamma}{\partial x^1}, \ldots, \frac{\partial \gamma}{\partial x^m}, \gamma(x)\right) = 0 \tag{9.1}$$

$$x = (x_1, \ldots, x_n) \in R^m = X \quad . \tag{9.2}$$

(9.1) is an equation for a map

$$x \rightarrow \gamma(x) \in R^m = Y \quad . \tag{9.3}$$

F is a map with the following domain and range:

$$F: R^{(n+1)\,m} \rightarrow R^p \quad . \tag{9.4}$$

Set:

$$V_1 = \frac{\partial}{\partial x^1}, \ldots, V_n = \frac{\partial}{\partial x^n} \quad . \tag{9.5}$$

$V_1, \ldots, V_n$ forms an abelian Lie algebra of vector fields on X. We can rewrite (9.1) in the following form:

$$F(V_1(\gamma), \ldots, V_n(\gamma), \gamma) = 0 \quad . \tag{9.6}$$

Let us apply the Euler discretization process sketched in Section 8. We can do this in even greater generality.

Suppose that X is a general manifold, and $V_1, \ldots, V_n$ are vector fields on X that generate one-parameter groups

$$t \to g_i(t): X \to X \quad ,$$
$$1 \le i, j \le n \quad ,$$

of diffeomorphisms. Let Y remain as a real vector space, so that we can add values of maps whose image lies in Y. Then, we have:

$$V_i(\gamma)(x) = \lim_{h \to 0} \left(\gamma(g_i(h)x) - \gamma(x) \right) h^{-1} \quad . \tag{9.7}$$

To Euler-discretize the differential equation (9.6), we can replace $V_i(\gamma)$ with the maps $V_i(h_i)$ defined as follows:

$$V_i(h_i)(\gamma)(x) = h_i^{-1} \left(\gamma(g_i(h_i)x) - \gamma(x) \right) \tag{9.8}$$
$$= h_i^{-1} \left(g_i(h_i)^*(\gamma)(x) - \gamma(x) \right)$$
$$= h_i^{-1} \left(g_i(h_i)^*(\gamma) - \gamma \right)(x) \quad . \tag{9.9}$$

Hence,

$$V_i(h_i)(\gamma) = h_i^{-1} \left(g_i(h_i)^*(\gamma) - \gamma \right) \quad . \tag{9.10}$$

We can associate with the differential equation (9.6) the following differential operator:

$$D\,(\gamma) = F\,(V_1\,(\gamma), \, \dots \, , V_n\,(\gamma), \gamma) \quad . \tag{9.11}$$

Now, we can define the **Euler-discretized** operator

$$D\,(h_1, \, \dots \, , h_n) \tag{9.12}$$

by the following formula:

$$D\,(h_1, \, \dots \, , h_n)\,(\gamma) = F\,\big(V_1\,(h_1)\,(\gamma), \, \dots \, , V_n\,(h_n)\,(\gamma), \gamma\big) \quad . \tag{9.13}$$

Now, for the purpose of constructing approximations to the solutions of the differential equation system (9.6), it is not necessary to solve the associated difference equation

$$D\,(h_1, \, \dots \, , h_n)\,(\gamma)\,(x) = 0 \tag{9.14}$$

for **all** $x \in X$, but only for certain subsets. Let us formulate this in the following way:

Suppose that, for each n-tuple

$$(h_1, \, \dots \, , h_n)$$

belonging to a neighborhood of O in R^n, we are given a subset

$$X\,(h_1, \, \dots \, , h_n) \subset X \quad . \tag{9.15}$$

We solve the difference equation (9.13), for each $x \in X\,(h_1, \, \dots \, , h_n)$. The resulting family of maps

$$\gamma\,(h_1, \ldots , h_n):\ X\,(h_1, \ldots , h_n) \to Y$$

then is to provide approximate solutions of the differential equation

$$D\,(\gamma) = 0 \quad .$$

Having formulated the process of discretization of a differential system, we are in position to ask the following question:

Which differential systems can be discretized in such a way that the approximating difference system can be solved in terms of primitive recursive functions?

Let us now examine several of the simplest examples of differential systems from this point of view.

10. RECURSIVE DISCRETIZATION OF A SINGLE FIRST-ORDER PARTIAL DIFFERENTIAL EQUATION FOR A SCALAR-VALUED FUNCTION

Consider a partial differential equation of the following form:

$$F\,(x, \gamma, V_1\,(\gamma), \ldots , V_n\,(\gamma)) = 0 \quad , \tag{10.1}$$

with

$$V_1 = \frac{\partial}{\partial x_1}, \ldots , V_n = \frac{\partial}{\partial x_n} \quad , \tag{10.2}$$

$$\gamma:\ R^n \to R \quad , \tag{10.3}$$

$$F:\ R^{m+1} \to R \quad . \tag{10.4}$$

This is the traditional example of a partial differential equation which is "integrable" in terms of ordinary differential equations via the theory of **Cauchy characteristics**.

An important special case is the **Hamilton-Jacobi equation** of mathematical physics

TO BE CONTINUED

11. LINEAR ELEMENTARY RECURSIVE FUNCTIONS

As we have seen, the theory of recursive functions deals with solutions of difference equations of the following form:

$$\gamma\,(x,\,n+1) = f\,(x,\,n,\,\gamma\,(x,\,n)) \qquad\qquad (11.1)$$
$$\text{for } x \in X,\, n \in N \quad.$$

A solution

$$(x,\,n) \to \gamma\,(x,\,n)$$

of (11.1) is a map

$$\gamma\colon\ X \times N \to Y \quad. \qquad\qquad (11.2)$$

X and Y are sets given in advance. (Of course, we might want X and Y to vary also at some point.)

f is a map with the following domain and range:

$$f\colon\ X \times N \times Y \to Y \quad. \qquad\qquad (11.3)$$

From the strict recursive function point of view, f should also be defined recursively, or in terms of other previously defined functions and admissible operators.

Definition. The primitive recursive system (11.1) is **linear** if the following conditions are satisfied:

a) Y is a commutative semigroup, with the semigroup operation
   denoted additively. $\qquad\qquad (11.1)$

b) If $\gamma_1, \gamma_2 \colon X \times N \to Y$ are solutions of (8.1), so is their
sum

$$\gamma = \gamma_1 + \gamma_2 \tag{11.2}$$

with

$$\gamma\,(x,\,n) = \gamma_1\,(x,\,n) + \gamma_2\,(x,\,n)$$
for $x \in X,\ n \in N$.

12. THE GENERALIZED JET SPACES ASSOCIATED WITH VECTOR FIELD SYSTEMS

Let X be a C^∞ manifold, and let

$$V_i \quad , \tag{12.1}$$

$$1 \le i,\,j,\,\ldots \le n \tag{12.2}$$

be a collection of vector fields on X. Let $V\,(X)$ be the set of all C^∞ vector fields on X.

$V\,(X)$ is both a Lie algebra — under the Jacobi - Lie bracket operation — and a module over the ring

$$\mathbf{F}\,(X)$$

of C^∞, real-valued functions on X. We can associate with the collection $V_1, \ldots, V_n$ the $\mathbf{F}\,(X)$-submodule V of $V\,(X)$ that it generates, and that we denote as follows:

$$\{V_1, \ldots, V_n\}_{\mathbf{F}\,(X)} \quad \cdot \tag{12.3}$$

Let Y be a real, finite dimensional, vector space.

One of the main themes in this paper is the study of differential operators on X which are "functions" of the first-order operators (12.1), i.e., of the following form:

$$D (\gamma) = F (x, V_i, V_i V_j, \dots) (\gamma) (x) \qquad (12.4)$$
$$\text{for } \gamma \text{ a map: } X \to Y \quad .$$

Let us denote the set of these differential operators as follows:

$$\mathbf{D} (V_1, \dots , V_n) \quad . \qquad (12.5)$$

In the partial differential equation literature, the **Hormander   Laplacian**

$$D = V_1 V_1 + \dots + V_n V_n \qquad (12.6)$$

is the most famous example. (Hormander's work linked up the analytic property of **hypoellipticity** with the geometric/control-theoretic property of **controllability**.)

In this section, I will present the jet-theoretic, "kinematic" foundation for the study of differential operators of the form (12.4).

Theorem   12.1. Let $V_1', \dots , V_m'$ be a collection of vector fields on X that belong to the submodule $\{ V_1, \dots , V_n \}_{\mathbf{F} (X)}$, i.e., are such that

$$\{ V_1', \dots , V_m' \}_{\mathbf{F} (X)} \subseteq \{ V_1, \dots , V_n \}_{\mathbf{F} (X)} \quad . \qquad (12.7)$$

Then,

$$\mathbf{D} (V_1', \dots , V_m') \subseteq \mathbf{D} (V_1, \dots , V_n) \quad . \qquad (12.8)$$

Proof. The hypothesis (12.7) means that there are relations of the following form:

$$V_1' = f_{11} \, V_1 + \cdots + f_{1n} \, V_n$$

$$\vdots \tag{12.9}$$

$$V_m' = f_{m1} \, V_1 + \cdots + f_{mn} \, V_n \quad ,$$

where the f's form an m x m matrix of functions on X.

If D is an element of $\mathbf{D}\,(V_1', \ldots, V_m')$, then it can be written in the following form:

$$D = F\,(x,\, V_a'',\, V_a'\, V_b',\, \ldots) \tag{12.10}$$
$$1 \le a, b \le m \quad ,$$

where "F" is a smooth function of the appropriate number of variables.

Now, if (12.9) is used to express the V_a' in terms of the V_i, the iterated operators

$$V_a',\; V_a'\, V_b',\; V_a'\, V_b'\, V_c',\; \ldots \tag{12.11}$$

can be expressed in terms of the

$$V_i,\; V_i\, V_j,\; V_i\, V_j\, V_k,\; \ldots \tag{12.12}$$

with elements of $\mathbf{F}\,(X)$ as coefficients. This proves (12.8).

Q.E.D.

Let us now use these ideas to generalize the jet space construction. Let

$$\Gamma\,(X, Y) = \{\gamma \colon X \to Y\}$$

again be the set of smooth maps from X to Y.

Definition. Let **V** be a subset of **V** (X), the set of smooth vector fields on X. Let r be a positive integer. Introduce the following equivalence relation in the set Γ (X, Y) x X:

$$(\gamma, x) \sim (\gamma', x') \quad \text{if and only if}$$
$$x = x' \quad . \tag{12.13}$$

For each r-tuple $V_1, \ldots, V_r \in$ **V** ,
$$V_1 (\gamma) (x) = V_1 (\gamma') (x)$$
$$V_1 V_2 (\gamma) (x) = V_1 V_2 (\gamma') (x)$$
$$\begin{array}{c} \bullet \\ \bullet \\ \bullet \end{array} \tag{12.14}$$
$$V_1 \ldots V_r (\gamma) (x) = V_1 \ldots V_r (\gamma') (x) \quad .$$

The quotient of Γ (X, Y) x X under this equivalence relation is denoted by

$$J^r (X, Y, V) \quad ,$$

and called the space of **r-jets of maps:** **X** $\to$ **Y relative to V.**

Theorem 12.2. Let **V** and **V**' be subsets of **V** (X) satisfying the following relation:

$$\mathbf{V} \subset \mathbf{V}' \quad . \tag{12.15}$$

Then, each equivalence class of type (12.12) — (12.13) relative to **V**' is contained in an equivalence class relative to **V**. The inclusion of equivalence classes then induces a map

$$J^r (X, Y, V') \to J^r (X, Y, V) \quad . \tag{12.16}$$

Theorem 12.3. Keep the notation and hypotheses of Theorem 12.1. Suppose, in addition, that the following conditions are satisfied:

$$\mathbf{V'} \text{ is an } \mathbf{F}(X)\text{-submodule of } \mathbf{V}(X) \tag{12.17}$$

$$\mathbf{V'} \text{ is the smallest } \mathbf{F}(X)\text{-submodule of } \mathbf{V}(X)$$
$$\text{which contains } \mathbf{V} \quad . \tag{12.18}$$

Then, the equivalence relations (12.12) — (12.13) defined relative to $\mathbf{V}$ and $\mathbf{V'}$ are the same, i.e.,

$$J^r(X, Y, \mathbf{V}) = J^r(X, Y, \mathbf{V'}) \quad . \tag{12.19}$$

13. DISCRETE-TIME FEEDBACK CONTROL FROM THE RECURSIVE FUNCTION POINT OF VIEW

Many of the standard recursive algorithms of computer program theory could also be considered as **closed loop feed-back control systems**. The aim of this section is to investigate when, conversely, feedback control systems can be considered as **recursive systems**.

Consider a **feedback control system** of the following difference equation form:

$$x(n+1) = f(n, x(n), u) \tag{13.1}$$
$$\text{for each } n \in N \quad , \quad x(n) \in X \quad .$$

X is the **state space**; $u \in U$, the **control space**.

$$f: N \times X \times U \rightarrow X \quad . \tag{13.2}$$

In the most elementary engineering situations, "f" is given, and the problem is to choose the control "u" to achieve some goal for $x(n)$; for example, for large values of n, to have $x(n)$ follow another discrete path

$$n \rightarrow x_o(n)$$

in X.

The simplest set-up is to assume that the control "u" is determined in terms of the state x by a relation of the following form, called **feedback**:

$$u\ (n) = K\ (n,\ x\ (n)) \tag{13.3}$$

$$K\colon\ N \times X \to U \quad . \tag{13.4}$$

If (13.3) is introduced into (13.2), the following difference equation is obtained:

$$x\ (n + 1) = f\big(n,\ x\ (n);\ K\ (n,\ x\ (n))\big) \quad . \tag{13.4}$$

From the more general "differential system" point of view, (13.1) and (13.3) form a subsystem of the originally given system (13.1). The problem is:

**To choose such a subsystem so as to achieve
a given subject to side constraints of "stability" and "robustness."** (13.5)

This is a typical "engineering" point of view.

From the computer science/ recursive function point of view, we recognize (13.3) — (13.4) as defining n → (x (n), u (n)) a primitive recursive function, providing that "f" and "K" already belong to the class of primitive recursive functions. From the traditional engineering point of view, "f" and "K" are to be regarded as "known" only approximately.

The point of view of adaptive control theory is somewhere in between the traditional engineering and recursive function points of view. Intuitively, one wants to build up "f" and "K" by "learning." Mathematically, additive control theory uses Liapounov function methodology in a very strong way: I believe that such Liapounov method-ology would also be very useful in computer programming theory.

As an example of a "recursive" feedback control system, let us add to (13.1) and (13.3) a primitive recursive definition of "f" and "K;" say, relations of the following form:

$$f (n + 1, x, u)$$
$$= F \left(n, x, u, f (n, x (u))\right) \qquad (13.6)$$

$$K (n + 1, x, u)$$
$$= G \left(n, x, u, K (n, x (u))\right) \quad . \qquad (13.7)$$

14. LINEAR CONTROL SYSTEMS

Let us now specialize the material of Section 13 to the case of the linear feedback control systems. One such case is of the following form:

$$x (n + 1) = A (n) (x (n)) + B (n) (u (n)) \quad , \qquad (14.1)$$

$$x (n) \in X \quad , \qquad (14.2)$$

$$u (n) \in U \quad . \qquad (14.3)$$

X and U are vector spaces over a given ground field

$$A (n) \in L (X, X) = \text{vector space of}$$
$$\text{linear maps:} \ X \to X \quad , \qquad (14.4)$$

$$B (n) \in L (U, X) \quad . \qquad (14.5)$$

In the engineering applications, the field over which such vector spaces are defined is the real or complex numbers. The data is usually given as matrices over this field in a more general and mathematical direction. One can consider the notion of "field" replaced by a general algebraic notion, such as a ring or commutative semigroup.

A **linear feedback control law** for the linear system (14.1) may be defined by a relation of the following form between "input" and "state:"

$$u\ (n) = K\ (n)\ (x\ (n))\quad ,\qquad\qquad (14.6)$$

$$K\ (n) \in L\ (X, U)\quad .\qquad\qquad (14.7)$$

Theorem 14.1. Suppose given a linear control system of the form (14.1) — (14.3). Let the "input" and "state" be related linearly via the feedback relation (14.7). Then, the state curve of the system

$$n \to x\ (n)$$

is determined recursively by the following difference equation:

$$x\ (n + 1) = [A\ (n) + B\ (n)\ K\ (n)]\ (x\ (n))\quad .\qquad\qquad (14.8)$$

Proof. (14.8) is obtained by substituting (14.6) into (14.1).

Q.E.D.

Remark. I have dignified with the name of "theorem" the "trivial" argument which leads from (14.1) and (14.6) to (14.8) of this possible "categorical" importance. In this case, we are working within the category of **linear** maps between vector spaces, and maps from **N** to that category.

The argument going from {14.1, 14.6} to {14.8} essentially involves a morphism of this category of linear maps, namely

$$\{K\ (n)\} \times \{A\ (n), B\ (n)\} \to \{A\ (n) + B\ (n)\ K\ (n)\}\quad .\qquad\qquad (14.9)$$

In certain cases, C. Martin and I have used the algebraic - geometric properties of the map (14.9) to prove results of an important control-theoretic nature about the "closed loop" system (14.8), such as the possibility of choosing n $\to$ K (n) to make the system (14.8) asymptotically stable.

In the extremely important 1950's work of Kalman on the linear regulator problem, the key question is the conditions for the existence of a quadratic Liapounov function for the closed loop linear system (14.8). Let us briefly examine what is involved algebraically.

Definition. A **quadratic form** on the vector space X may be defined as a map

$$\beta: X \times X \to \text{(ground field)}$$
$$(x_1, x_2) \to \beta(x_1, x_2) \quad , \qquad (14.10)$$

such that the following conditions are satisfied:

β is a bilinear map relative to the vector space
structure on X   .   (14.11)
$$\beta(x_1, x_2) = \beta(x_2, x_1) \qquad (14.12)$$
for $x_1, x_2 \in X$.

Traditionally, the set of such symmetric, quadratic forms is identified with the set of symmetric matrices. This can be formulated algebraically as follows.

Suppose that the vector space X has a fixed symmetric, bilinear form that we denote as follows:

$$(x_1, x_2) \to x_1 \cdot x_2 \quad . \qquad (14.16)$$

Remark. Think of $X = R^n$, and (14.16) as the usual "dot product" of vector analysis.

Definition. Assign to each $x_0 \in X$ the map

$$\alpha\,(x_0)\colon\; x \to x_0 \cdot x \quad . \tag{14.17}$$

$\alpha\,(x_0)$ is an element of the dual vector space to X. The map

$$x_0 \to \alpha\,(x_0) \tag{14.18}$$

is a linear map from X to its dual vector space. The given quadratic form (14.16) is said to be **strongly non-degenerate** if the following condition is satisfied:

> The map (14.18) is an isomorphism from the vector
> space X to its dual. $\hspace{3cm}$ (14.19)

Theorem 14.2. Suppose that the fixed quadratic form (14.6) is strongly non-degenerate. Given a quadratic form

$$\beta\colon\; X \times X \to \text{(ground field)} \quad , \tag{14.20}$$

there is a unique element

$$\alpha\,(\beta) \in L\,(X,\,X) \tag{14.21}$$

satisfying the following conditions:

$$\beta\,(x_1,\,x_2) = \alpha\,(\beta)\,(x_1) \cdot x_2 \tag{14.22}$$
$$\text{for } x_1,\,x_2 \in X \quad ,$$

$$\alpha\,(\beta)^T = \alpha\,(\beta) \quad , \tag{14.23}$$

where

$$\alpha \to \alpha^T \tag{14.24}$$

$$L\,(X,\,X) \to L\,(X,\,X) \tag{14.25}$$

is the transpose map on linear maps (classically, on "matrices") associated with the quadratic form (14.6).

Proof. The existence of $\alpha\,(\beta)\,(x_1)$ satisfying (14.22) follows from the strong non-degeneracy hypothesis. $x \rightarrow \alpha\,(\beta)\,(x)$ as a linear map can be readily proved from the bilinearity of β. The "transpose" operation (14.25) is defined by the following relation:

$$\alpha\,(x_1) \cdot x_2 = x_1 \cdot \alpha^T\,(x_2) \qquad (14.26)$$
$$\text{for } x_1, x_2 \in X \quad .$$

(14.23) follows from (14.25).

Q.E.D.

Given a map

$$n \rightarrow \beta\,(n) \qquad (14.27)$$

of **N** into the set of such quadratic forms, and a solution

$$n \rightarrow x\,(n)$$

of the linear difference equation

$$x\,(n + 1) = A(n)\,x\,(n) \quad , \qquad (14.28)$$

set:

$$E\,(n) = \beta\,(n)(x\,(n),\, x\,(n)) \quad . \qquad (14.29)$$

Remark. Think of "E" as suggesting **energy**.

For each n, let $\alpha\,(n)$ be the linear map
$$\alpha\,(n)\colon X \rightarrow X \qquad (14.30)$$

satisfying the following conditions:

$$\beta (n) (x_1, x_2) = \alpha (n) (x_1) \cdot x_2 \tag{14.31}$$
for $x_1, x_2 \in X$

$$\alpha (n)^T = \alpha (n) \quad . \tag{14.32}$$

We can then write (14.28) as follows:

$$E (n) = \alpha (n) (x (n)) \cdot x (n) \quad . \tag{14.33}$$

Let us now calculate how $E (n)$ changes with n as $x (n)$ evolves via the discrete dynamical system (14.27):

$$\begin{aligned}
E (n+1) &= \alpha (n + 1) x (n + 1) \cdot x (n + 1) \\
&= \alpha (n + 1) A (n) x (n) \cdot A (n) x (n) \\
&= (A (n)^T \alpha (n + 1) A (n)) (x (n)) \cdot x (n) \quad .
\end{aligned} \tag{14.34}$$

Theorem 14.3. Suppose that $n \to \alpha (n)$ satisfies the following difference equation

$$\alpha (n) = A (n)^T \alpha (n + 1) A (n) \quad . \tag{14.35}$$

Then,

$$E (n + 1) = E (n) \tag{14.36}$$

for all $n \in \mathbf{N}$, all solutions

$$n \to x (n)$$

of the discrete-time, linear dynamical system (14.27)

Proof. Follows from (14.33).

Q.E.D.

Let us now see how "equality" in (14.35) can be replaced by "inequality."

Theorem 14.4. Suppose that the ground field of the vector space is a sub-field of the real numbers, so that it inherits the usual total ordering "L" of the real numbers.

Introduce a partial ordering on the vector space of symmetric linear maps: $X \to X$:

$$\alpha_1 \leq \alpha_2 \text{ if and only if}$$
$$\alpha_1 (x) \cdot x \leq \alpha_2 (x) \cdot x \tag{14.36}$$
$$\text{for all } x \in X .$$

Suppose that
$$A (n)^T \alpha (n + 1) A (n) \leq \alpha (n) . \tag{14.37}$$

Then,

$$E (n + 1) \leq E (n) . \tag{14.38}$$

Proof. Follows from (14.32), (14.33), and (14.36).

Q.E.D.

15. EQUIVALENCE FOR DIFFERENCE AND DIFFERENTIAL SYSTEMS IN THE SENSE OF DIFFERENTIAL GEOMETRY AND RECURSIVE FUNCTION THEORY

For over one hundred years, mathematicians have learned that it is instructive to study "morphisms" of mathematical structures. Sophus Lie was the first to systematically develop and use this point of view in the theory of differential systems.

In a magnificent series of papers from about 1895 – 1915, Elie Cartan described algorithms for the effective realization of Lie's ideas across a wide spectrum of examples. Lie and Cartan's work was isolated from the mainstream of mathematics until the 1950's: To this day, understanding and implementing their algorithms is a challenge at the very limit of the computational and conceptual ability of the differential geometric community.

In his influential book on recursive functions, Hartley Rogers describes a program of "recursive equivalence" for recursive functions that amounts to an analogy with the ideas of Lie and Cartan. However, so far as I can see, Rogers' equivalence program is not very extensively developed in his book, or in the subsequent literature on the theory of recursive functions.

One of the great features of the 1950's Ehresmann theory of sets of mappings is that it put the Lie – Cartan equivalence theory on a much firmer conceptual foundation. (However, it did not really make the algorithmic problems simpler!) Let us then use such a formulation.

Let X and Y be manifolds, r a positive integer, with

$$J^r(X, Y)$$

the manifold of r-jets of smooth maps: $X \to Y$. Let

$$DS \subset J^r(X, Y) \tag{15.1}$$

define a differential system. A map $\gamma: X \to Y$ is a **solution** if

$$j^r(\gamma(x)) \subset DS \tag{15.2}$$

Let

$$(X, Y) \; DS' \subset J^r (X, Y) \tag{15.3}$$

define another differential system. A map

$$\phi: J^r (X, Y) \to J^r (X, Y') \tag{15.4}$$

is an **equivalence** from differential system (15.1) to differential system (15.4) if:

$$\phi (DS) \subset DS' \quad . \tag{15.5}$$

(15.5) defines a system of differential equations for ϕ. The idea of "recursiveness" is to find the conditions for the existence of such a ϕ which belongs to a previously described class of maps.

Let us consider some simple examples. Consider a first-order ordinary differential equation of the following form:

$$\frac{dx}{dt} = f (x, \; t) \tag{15.6}$$
$$x \in R \quad .$$

Let us find the conditions of equivalence of (15.6) with the standard model of a first-order ordinary differential equation:

$$\frac{dy}{dt} = 1 \quad . \tag{15.7}$$

Suppose $t \to y (t)$ is a solution of (15.7), and

$$x (t) = g(y(t), \; t) \quad . \tag{15.8}$$

Then

$$\frac{dx}{dt} = g_x \frac{dy}{dt} + g_t$$

$$= g_x \, (y \, (t), \, t) + g_t \, (y \, (t), \, t) \quad . \tag{15.9}$$

(Subscripts denote partial derivatives.)

Then, $t \to x \, (t)$ given by (15.8) is a solution of (15.6) if and only if:

$$f\big(g(y \, (t), \, t), \, t\big)$$
$$= g_x \, (y \, (t), \, t) + g_t \, (y \, (t), \, t) \quad . \tag{15.10}$$

We have proved:

Theorem 15.1. The map (15.8) sends every solution of (15.7) into a solution of (15.6) if and only if the function g satisfies the following partial differential equation:

$$g_x \, (y, \, t) + g_t \, (y, \, t) = f \, (g \, (y, \, t), \, t) \quad . \tag{15.11}$$

Proof. Substitute (15.8) into (15.6), and use (15.10).

Q.E.D.

Conversely, there is a classical algorithm constructing a solution of (15.11) from the "general solution" of the ordinary differential equation (15.6). Since this involves inverting maps defined via the "general solution," we see that, in this case, the analogue of "recursive equivalence" is satisfied.

RECURSIVE SYSTEMS AND JET BUNDLES

1. INTRODUCTION

The theory of differential systems and jet bundles - as it evolved from the pioneering work of the 1950's and early 60's - became very abstract, temporarily precluding its its extensive utilization for applications. Since I believe that it is also very important for 'applied' purposes - and in a wide variety of disciplines, from Computer Science through Control Theory to Elementary Particle Physics - I will continue here exposition of some of ts principles in a 'geometric' form that might be more accessible to its potential 'applied' audience, as I have done for the physical sciences in some of my books .

Ever since their introduction in the 1950's, the Ehresmann jet bundles have been difficult to work with on a computational level. Although they have their origin in the relatively concrete 19th century work on systems of nonlinear partial differential equations, the computational aspects of the formalism have not been emphasized in the work since the 1950's. Since I am now proposing developing an extension of the Ehresmann theory to serve the 'computational' purposes of present-day computer science, it seems essential to refine the algebraic computational tools required to 'apply' the theory.

It is customary to do calculations on the jet bundles by using special coordinate systems adapted to the structure. While I have followed this course in my own mathematical physics books utilizing the Ehresmann ideas, I believe that it is possible to work more efficiently and algebraically by analyzing more carefully the algebraic structure of the differential forms on the manifold sttructures associated with the jet spaces. I will begin the development of such a formalism in this paper. As preparation for generalization to the systems of equations encountered in Computer Science, I will briefly describe the overall category-theoretic structure of the Ehresmann jet bundles, and then sketch parallel ideas of differential system theory.

2. A CATEGORY-THEORETIC DESCRIPTION OF THE EHRESMANN JET-BUNDLES AND DIFFERENTIAL SYSTEMS

Definitions. Let X and Y be finite dimensional, infinitely-differentiable, paracompact manifolds. Let $\Gamma(X, Y)$ be the set of smooth (i.e. infinitely-differentiable maps) from X to Y. Let **N** be the following category:

$$\text{The objects of } \mathbf{N} \text{ are the non-negative integers } \mathbf{N} \qquad (2.1)$$

$$\text{The morphisms of } \mathbf{N} \text{ are the pairs } (j, k) \ \varepsilon \ \mathbf{N} \ x \ \mathbf{N},$$
$$\text{such that: } j \leq k. \qquad (2.2)$$

Let **DF** (which stands for 'differentiable fiber') be the category whose objects are the finite dimensional, infinitely-differentiable, paracompact manifolds and whose morphisms are the infinitely differentiable, local-product fiber maps between such manifolds.

Theorem 2.1. Let X and Y be finite dimensional, infinitely-differentiable, paracompact manifolds. There is a contravariant functor:

$$J(X, Y)\colon \mathbf{N} \ \text{-->} \ \mathbf{DF} \qquad (2.3)$$

$J(X, Y)$ assigns to each object j of the category **N** a manifold $J^j(X, Y)$, called the **manifold of j-th order jets of mappings of X to Y**, and it assigns to each morphism (j, k) of the category **N**, a local-product, smooth, fiber map:

$$\pi(k, j)\colon J^k(X, Y) \ \text{-->} \ J^j(X, Y), \qquad (2.4)$$

such that, whenever $j \leq k$ and $k \leq l$,

$$\pi(l, j) = \pi(k, j)\pi(l, k) \qquad (2.5)$$

Further, there is a space $J^\infty(X, Y)$, called the space of **infinite order jets,** and a family of maps:

$$\pi(j): J^\infty(X, Y) \dashrightarrow J^j(X, Y) \tag{2.6}$$

such that, whenever $j \le k$,

$$\pi(j) = \pi(k, j)\pi(k) \tag{2.7}$$

For each $\gamma \in \Gamma(X, Y)$, each $k \in N$, there is a map:

$$j^k(\gamma): X \dashrightarrow J^k(X, Y) \tag{2.8}$$

called the **prolongation** or **k-jet** of γ . The following condition is satisfied:

$$\pi(k, j)j^k(\gamma) = j^j(\gamma) \quad \text{for } j \le k \tag{2.9}$$

In language of the Theory of Categories [], $J(X, Y)$ is a contravariant functor from the category **N** associated with the ordered set of non-negative integers to the category **SET** of sets and maps. The collection of maps 2.6 defines $J^\infty(X, Y)$ as a 'colimit' of this functor.

Proof. The $J^k(X, Y)$ and $J^\infty(X, Y)$ are defined as the quotients of $X \times \Gamma(X, Y)$ with respect to the following equivalence relations:

$$(x, \gamma) \approx^k (x', \gamma') \text{ iff.}$$
$$x = x'; \gamma \text{ and } \gamma' \text{ agree to the k-th order at x.} \tag{2.10}$$
$$\text{in local coordinate systems about x.}$$

$$(x, \gamma) \approx^\infty (x', \gamma') \text{ iff.}$$
$$x = x'; \gamma \text{ and } \gamma' \text{ agree to infinite order at x.} \tag{2.11}$$
$$\text{in local coordinate systems about x.}$$

The equivalence classes to which the pair (x, γ) belong is denoted as follows:

$$j^k(\gamma)(x) \qquad (2.12)$$

As x varies, we obtain a map:

$$j^k(\gamma): X \dashrightarrow J^k(X, Y) \qquad (2.13).$$

The projection maps $\pi(k, j)$ and $\pi(k)$ are obtained using the obvious inclusion relations beteen these equivalence relations. The manifold structures of the set $\{J^j(X, Y)\}$ and the local-product nature of the maps $\{\pi(k, j)\}$ are exhibited using local-coordinate patches for X and Y.

Q.E.D.

Let us now see how the bundle $J^\infty(X, Y)$ is algebraically constructed from the family $\{ J^k(X, Y): k \ \epsilon \ N\}$ of bundles.

Theorem 2.2. The equivalence relation $\approx^\infty$ defined by 2.9 on the set $X \times \Gamma(X, Y)$ is the intersection of the family $\{\approx^k: k \ \epsilon \ N\}$ of equivalence relations. In other words, $\approx^\infty$, as a subset of the Cartesian product of $X \times \Gamma(X, Y)$ with itself, is the intersection of the family of subsets assocated with the equivalence relations $\{\approx^k: k \ \epsilon \ N\}$.

Proof. Geometrically, two smooth maps between manifolds agree to 'infinite order' at a point if and only if they meet to all finite orders. Theorem 2.2 now follows from 2.8 and 2.9.

Q.E.D.

We can now present the definition of a 'differential system':

Definition. A subset DS $\subset J^k(X, Y)$ is said to define a **differential system of order k of maps from X to Y**. A $\gamma \ \epsilon \ \Gamma(X, Y)$ is said to be a **solution** of DS if the following condition is satisfied:

$$\text{For each } x \ \epsilon \ X, \ j^k(\gamma)(x) \ \epsilon \ DS \qquad (2.14)$$

The collection of such solutions is denoted as:

$$\Gamma(DS) \tag{2.15}$$

Let us now pursue an algebraic version of some of this material.

3. DIRECTED SETS OF EQUIVALENCE RELATIONS.

The construction of the jet-bundles described in Section 2 has a general feature that will be used as a basis for generalization to systems of recursion equations. The main property that I want to capture is the structure of the infinite-order jet bundle as a sort of 'limit' of the sequence of finite-order jet bundles. In category theory, this is done by algebracizing and generalizing the notions of 'limits'. Here, I will only deal with limits of categories consisting of 'equivalence relations', where the relevant 'limit' notions can be described more concretely.

Definition.   Let Z be a set. A **(binary) relation** on Z is a subset of Z x Z. The **relation** **R** is **reflexive** iff:

$$(z, z) \; \varepsilon \; \mathbf{R} \text{ for } z \; \varepsilon \; Z; \tag{3.1}$$

transitive iff.

$$(z, z') \; \varepsilon \; \mathbf{R} \text{ and } (z', z'') \; \varepsilon \; \mathbf{R} \text{ implies } (z, z'') \; \varepsilon \; \mathbf{R}; \tag{3.2}$$

and **symmetric** iff.

$$(z, z') \; \varepsilon \; \mathbf{R} \text{ implies } (z', z) \; \varepsilon \; \mathbf{R}. \tag{3.3}$$

Let us first recall the definition of a 'pre-order' and 'directed set', which play a basic role in the geometric applications of category theory, [].

Definition Let Z be a set. A **pre-order** on Z is a relation **R** on Z that is transitive and symmetric. If **R** is a pre-order on Z, it is customary to use the following notation:

$$\text{"}(z, z') \; \varepsilon \; \mathbf{R}\text{" is written as "} z \leq z' \text{ "} \tag{3.4}$$

Such a pre-order structure is said to define Z as a **directed** **set** if satisfies the following condition:

If z and z' are elements of Z, then there is a z" ε Z such that:

$$z \leq z'' \text{ and } z \leq z'' \tag{3.5}$$

Let us now list some basic definitions cenering around the concept of 'equivalence relation'.

Definition.   The relation $\mathbf{R} \subset Z \times Z$ is an **equivalence** **relation** if it is reflexive, transitive and symmetric. The **symmetric** **part** of the relation $\mathbf{R} \subset Z \times Z$ is the relation $\mathbf{R'}$ defined as follows:

$$\mathbf{R}' = \{(z', z): (z, z') \text{ and } (z', z) \ \varepsilon \ \mathbf{R}\} \tag{3.6}$$

The **transitive closure of the relation** $\mathbf{R} \subset Z \times Z$ is the relation $\mathbf{R}''$ defined as follows:

$$\mathbf{R}'' = \{(z, z'): \text{The exists an integer n and an n-tuple} \tag{3.7}$$
$$(z_1, ..., z_n) \text{ of elements of } Z \text{ such that:}$$
$$(z, z_1), (z_1, z_2), ..., (z_n, z') \ \varepsilon \ \mathbf{R}\}$$

Theorem 3.1. Let $R \subset Z \times Z$ be a relation on the set Z and let R'' be the transitive closure relation defined by formula 3.5. Then, R'' is transitive. In particular, R'' is the intersection of all subsets of $Z \times Z$ which are transitive relations and contain R .

Proof. Let (z, z') and $(z', z'') \in R''$. Then there exists integers n and m, an n-tuple $(z_1, ..., z_n)$, and an m-tuple $(y_1, ..., y_m)$, of elements of Z such that:

$$(z, z_1), (z_1, z_2), ..., (z_n, z') \in R \qquad (3.8)$$

$$(z', y_1), (y_1, y_2), ..., (y_m, z'') \in R \qquad (3.9)$$

Putting these two chains together provides a chain:

$$(z, z_1), (z_1, z_2), ..., (z_n, z'), (z', y_1), (y_1, y_2), ..., (y_m, z'') \qquad (3.10)$$

of elements of R connecting z to z", completing the proof that R'' is transitive.
Q.E.D.

Theorem 3.2. Let $R \subset Z \times Z$ be a binary relation on the set Z such that R contains the diagonal subset of $Z \times Z$. Let R' be the transitive closure of R, and let R'' be the symmetric part of R'. Then, R'' is an equivalence relation on Z

Proof. R'' is reflexive, since it contains the diagonal subset of $Z \times Z$, and is symmetric, since it is the symmetric part of a relation. To complete the proof, we must show that R'' is transitive.

Let (z, z') and $(z', z'') \in R''$. Since $R'' \subset R'$, and R' is transitive, we have:

$$(z, z'') \in R'. \qquad 3.11)$$

Since R'' is symmetric, we have:

$$(z', z) \text{ and } (z'', z') \in R'' \subset R' \qquad (3.12)$$

Transitivity of $\subset$ **R'** again implies that:

$$(z'', z) \; \varepsilon \; \mathbf{R'} \tag{3.13}$$

3.10 and 3.12 now imply that:

$$(z, z'') \; \varepsilon \; \mathbf{R''} \tag{3.14}$$

which completes the proof of Theorem 3.2..

Definition. Let **R** be a relation on the set Z. A map $\pi\colon Z \dashrightarrow Y$ is said to be a **projection map relative to R** or **to pass to the quotient relative to the relation R** if the following condition is satisfied:

$$\pi(z) = \pi(z') \text{ for all } (z, z') \; \varepsilon \; \mathbf{R} \tag{3.15}$$

Theorem 3.3. Let **R** be a reflexive relation on the set Z. Suppose that $\pi\colon Z \dashrightarrow Y$ is a map which is a projection map relative to **R**. Let **R''** be the symmetric part of the transitive closure **R'** of **R**. Then π is a projection map relative to **R'** and **R''**.

Proof. Let us suppose that π satisfies 3.4. Let $(z_1, \ldots, z_n)$ be an n-tuple of elements of Z such that:

$$(z, z_1), (z_1, z_2), \ldots, (z_n, z') \; \varepsilon \; \mathbf{R} \tag{3.16}$$

Applying 3.15 to each term of 3.16, we have:

$$\pi(z) = \pi(z_1) = \pi(z_2) = \ldots = = \pi(z') \tag{3.17}$$

This - together with the obvious fact that a projection map relative to a given relation is also a projection map relative to its symmetric part - implies that π is a projection map relative to **R'** and **R''**.

Q.E.D.

Definition. Let E be an equivalence relation on a set Z. For $z \in Z$, the **equivalence class of E containing z** is the subset:

$$\{z': (z, z') \in E \} \tag{3.18}$$

An **equivalence class of** E is a subset of Z which is the equivalence class of E of at least one element of Z. Each point of Z belongs to precisely one equivalence class. The **quotient of Z by** E, denoted as:

$$Z/E, \tag{3.19}$$

is the set of equivalence classes of E. The map:

$$\pi(E) : Z \dashrightarrow Z/E \tag{3.20}$$

which assigns to each $z \in Z$ the equivalence class to which it belongs is called the **canonical projection** or **quotient map associated with the equivalence relation** E.

Theorem 3.4. Let E be an equivalence relation on a set Z and let

$$\alpha: Z \dashrightarrow Y \tag{3.21}$$

be a map that is constant on the equivalence classes of E, i.e. that is a projection map relative to E. Then, there is a map:

$$\beta: Z/E \dashrightarrow Y \tag{3.22}$$

such that:

$$\alpha = \beta\pi(E) \, . \tag{3.23}$$

Conversely, such a map β exists, then α is constant on the equivalence classes of E.

Proof. Define β as follows:

$$\beta(\{z': (z, z') \in E \}) = \text{hypothesized common value} \qquad (3.24)$$
$$\text{of } \alpha \text{ on the equivalence class } 3.18$$

3.23 is then true by construction. The converse follows by reversing the argument.

Q.E.D.

Theorem 3.5. Let $\mathbf{E} \subset Z \times Z$ be a family of equivalence relations on the set Z. Let E be the intersection of all elements of $\mathbf{E}$, considered as subsets of $Z \times Z$. Then, E is also an equivalence relation on Z. In other words, the family of equivalence relations is closed under set-theoretic intersection.

Proof. The equations 3.1-3.3 which define 'equivalence relation' are - by inspection - satisfied by E if they are satisfied by all elements of $\mathbf{E}$.

Q.E.D.

Theorem 3.6 Let E and E' be two equivalence relations on Z, let:

$$\pi(E) : Z \dashrightarrow Z/E, \qquad (3.25)$$

$$\pi(E') : Z \dashrightarrow Z/E' \qquad (3.26)$$

be the quotient maps defined above. Suppose that the following condition is satisfied:

$$E' \subset E \qquad (3.27)$$

Then, there is a map:

$$\pi(E, E'): Z/E' \dashrightarrow Z/E \qquad (3.28)$$

such that:

$$\pi(E') = \pi(E, E')\pi(E) \tag{3.29}$$

Condition 3.25 can be described as the commutativity of the following diagram of mappings:

$$
\begin{array}{ccc}
Z \quad \text{(identity map)} \quad Z & & \\
\Downarrow \pi(E) & & \Downarrow \pi(E') \\
Z/E & \Rightarrow & Z/E' \\
& \pi(E, E') &
\end{array}
\tag{3.30}
$$

Proof. Let $z \; \varepsilon \; Z$. Then,

$$\pi(E')(z) = \{z' \; \varepsilon \; Z: (z, z') \; \varepsilon \; E'\} \tag{3.31}$$

In view of the hypothesis 3.23, we have:

$$\pi(E')(z) \subset \pi(E)(z) = \{z' \; \varepsilon \; Z: (z, z') \; \varepsilon \; E\} \tag{3.32}$$

We see that each equivalence class of E' is contained in precisely one equivalence class of E. Define $\pi(E, E')$ by the following property:

$$
\begin{aligned}
&\pi(E, E')(\text{an equivalence class of } E') = \\
&\qquad \text{the equivalence class of } E \text{ which contains it.}
\end{aligned}
\tag{3.33}
$$

With this definition, we see that condition 3.25 is satisfied.

Q.E.D.

Theorem 3.7. Let **E** be a set of equivalence relations on Z, partially ordered by inclusion. Consider this partially-ordered set as a category **C(E)** in the usual way. Let **SET** be the category whose objects are 'sets' and whose 'morphisms' are mappings between sets. Define a mapping

$$F : C(E) \; --> \; SET \tag{3.34}$$

as follows:

To each object E of $C(\mathbf{E})$ assign the quotient set Z/E $\qquad$ (3.35)

To each morphism of $C(\mathbf{E})$, i.e each pair (E, E') of elements of $\mathbf{E}$ such that:

$$E' \subset E, \qquad (3.36)$$

assign the map $\pi(E, E')$ defined above. Then, F is a contravariant functor between the categories $C(\mathbf{E})$ and **SET**.

Proof. Relation 3.26 is just the condition that F be a contravariant functor.
$$\mathbf{Q.E.D.}$$

We can consider the partial ordering '$\leq$' on $\mathbf{E}$ that is dual to that defined by the inclusion relation 3.31, i.e.

$$E \leq E' \text{ iff. } E' \subset E \qquad (3.37)$$

Theorem 3.8. Let $\mathbf{E}$ be the set of all equivalence relations on Z, with the partial ordering '$\leq$' defined by 3. 33 Then, $\mathbf{E}$ is a directed set relative to this partial ordering.

Proof. Let E and E' be two elements of $\mathbf{E}$. Set

$$E'' = E \cap E' \qquad (3.38)$$

Using 3.33, we see that:

$$E \leq E'' \text{ and } E' \leq E'', \qquad (3.39)$$

i.e. 3.5 is satisfied and $\mathbf{E}$ is a directed set relative to the partial ordering 3.33.
$$\mathbf{Q.E.D.}$$

We can express the resulting 'reversal of arrows' in the following way:

Theorem 3.9. Let **E** be a set of equivalence relations on Z, satisfying the following condition:

$$\text{If } E \text{ and } E' \, \varepsilon \, \mathbf{E} \text{ , then also } E \cap E' \, \varepsilon \, \mathbf{E} \qquad (3.40)$$

Consider **E** as a directed set by choosing the partial order relation 3.33. Let C(**E**)* denote the corresponding category.   Then,   the mappings 3.29-3.31 define F as a functor from C(**E**)* to **SET**.

4. AN EXTENSION OF THE JET-SPACE FORMALISM TO INCLUDE RECURSIVE EQUATION SYSTEMS.

In order to construct a 'geometric' theory of systems of recursive equations that - in a suitable categorical sense - 'extends' the Ehresmann theory of differential systems briefly described in Section 2, it is algebraically reasonable to investigate other equivalence relations on $X \times \Gamma(X, Y)$. We now define one class of such equivalence relations that - as we shall see at the end of this Section - is suitable for the definition of the 'classical' difference equations.

Definition Let X and Y be arbitrary sets and let $\Gamma(X, Y)$ be a collection of maps with domain X and range Y. Let:

$$\delta: X \dashrightarrow (\text{set of finite subsets of } X) \qquad 4.1)$$

be a map which assigns to each point of X a finite subset of X. Let:

$$\delta(X) = \text{set of all maps of form 4.1.} \qquad (4.2)$$

Now, choose a $\delta \, \varepsilon \, \delta(X)$ and introduce the following equivalence relation on $X \times \Gamma(X, Y)$:

$$(x, \gamma) \approx^{\delta} (x', \gamma') \text{ iff. the following conditions are satisfied:}$$

$$x = x';$$
(4.3)

γ and $\gamma\,'$ agree on the points of $\delta(x)$.
(4.4)

The quotient of $X \times \Gamma(X, Y)$ by this equivalence relations is denoted as:

$$J^\delta(X, Y)$$
(4.5)

If $(x, \gamma) \in X \times \Gamma(X, Y)$, we denote the equivalence class of the equivalence relation 4.3-4.4 to which (x, γ) belongs as follows:

$$j^\delta(\gamma)(x)$$
(4.6).

Let us define a partial ordering '$\leq$' on $\delta(X)$ as follows:

$$\delta \leq \delta' \text{ iff. } \delta(x) \subset \delta'(x) \text{ for all } x \in X.$$
(4.7)

Theorem 4.1. Let $\delta \in \delta(X)$. The Cartesian projection map:

$$(x, \gamma) \dashrightarrow x,$$
$$X \times \Gamma(X, Y) \dashrightarrow X$$
(4.8)

passes to the quotient to define a map:

$$J^\delta(X, Y) \dashrightarrow X.$$
(4.9)

For each $\gamma \in \Gamma(X, Y)$, there is a map:

$$j^\delta(\gamma): X \dashrightarrow J^\delta(X, Y)$$
(4.10)

such that, for $x \in X$, :

$$j^\delta(\gamma)(x) = \text{the element denoted by 4.6.}$$
(4.11)

$j^\delta(\gamma)$ is a cross-section map relative to themap 4.9, i.e. the map 4.10 followed by the map 4.9 is the identity map.

Proof. Define the map $j^\delta(\gamma)$ by 4.11, as x varies over X. The cross-section property follows.

$$\text{Q.E.D.}$$

Theorem 4.2. If δ and δ' are two elements of $\delta(X)$ such that $\delta \leq \delta'$, then there is a map:

$$\pi(\delta', \delta) \colon J^{\delta'}(X, Y) \dashrightarrow J^\delta(X, Y) \qquad (4.12)$$

with the following property: For each $\gamma \, \varepsilon \, \Gamma(X, Y)$,

$$\pi(\delta', \delta) \colon j^{\delta'}(\gamma)(x) = j^\delta(\gamma)(x) \qquad (4.13)$$

The map $\delta \dashrightarrow J^\delta(X, Y)$, $(\delta', \delta) \dashrightarrow \pi(\delta', \delta)$ defines a contavariant functor from the category associated with the partially-ordered set $\delta(X)$ to the category **SET**.

Proof. If 4.7 is satisfied, then:

$$(x, \gamma) \approx^{\delta'} (x', \gamma') \text{ implies } (x, \gamma) \approx^\delta (x', \gamma') \qquad (4.14)$$

i.e. the equivalence classes of the equivalence relation $\approx^{\delta'}$ are included in the equivalence classes of $\approx^\delta$. These inclusions induce maps on equivalence classes, which lead to the map 4.12 with the property 4.13. The 'contravariant functorial' property follows.

$$\text{Q.E.D.}$$

Having defined the set $J^\delta(X, Y)$ in this way as a generalization of the Ehresmann jet-bundle construction, we can extend the notion of 'solution':

Definition. Let γ be a map: $X \dashrightarrow Y$ that is an element of $\Gamma(X,Y)$. Let:

$$RS \subset J^\delta(X, Y) \qquad (4.15)$$

RS is said to define a **recursion system** with source X and target Y. γ is said to be a **solution** of the system RS iff:

$$j^{\delta'}(\gamma)(X) \subset RS .\qquad(4.16)$$

Example.   Classical Ordinary Difference Equations.

Suppose that:

$$X \text{ is the ring of integers.}\qquad(4.17)$$

$$Y \text{ is a set.}\qquad(4.18)$$

$$k \text{ is a non-negative integer}\qquad(4.19)$$

$$Y^k = \text{Cartesian product of } k \text{ copies of the set } Y\qquad(4.20)$$

$$F \text{ is a map: } X \times Y^{k+1} \longrightarrow R^m\qquad(4.21)$$

A **k-th order ordinary difference equation** associated with the above data is an equation of the form

$$F(x, y(x), y(x-1), ..., y(x-k)) = 0,\qquad(4.22)$$

$$\text{for } x \in X$$

A **solution** of the system 4.20 is a map:

$$X \longrightarrow Y, \ x \longrightarrow y(x)\qquad(4.23)$$

which satisfies 4.22 when substituted.

Theorem 4.3. Let $\delta \in \delta(X)$ be defined as follows:

$$\delta(x) = (x, x-1, ..., x-k). \tag{4.24}$$

Then,

$J^\delta(X, Y)$ may be identified with the Cartesian product $X \times Y^{k+1}$. (4.25)

Set:

DS = subset of $J^\delta(X, Y)$ corresponding,
under the identification 4.25, to the following subset

$$\{(x, x-1, ..., x-k) \in X \times Y^{k+1}: \ F(x, y(x), y(x-1), ..., y(x-k)) = 0. \tag{4.26}$$

of $X \times Y^{k+1}$.

Then, there is an identification between the solutions of the difference equation 4.22 and the set of $\gamma \in \Gamma(X, Y)$ such that:

$$j^\delta(\gamma)(X) \subset DS. \tag{4.27}$$

Proof. First, let us set up the identification of $J^\delta(X, Y)$ with $X \times Y^{k+1}$. Given

$$(x, \gamma) \in X \times \Gamma(X, Y), \tag{4.28}$$

assign to it the element:

$$(x, \gamma(x), \gamma(x-1), ..., \gamma(x-k)) \tag{4.29}$$

In this way, we obtain a map:

$$X \times \Gamma(X, Y) \dashrightarrow X \times Y^{k+1} \tag{4.30}$$

Let $E(\delta)$ be the equivalence relation on $X \times \Gamma(X, Y)$ such that:

$$J^\delta(X, Y) = X \times Y^{k+1}/E(\delta) \tag{4.31}$$

Then, the map 4. passes to the quotient relative to the equivalence relation E(δ) to define the map indicated in . It is readily seen that this map is one-one and onto, hence defines the required isomorphism.

THE EHRESMANN JET SPACES AND THE PROLONGATION CALCULUS

1. Introduction.

The geometry underlying the Ehresmann jet-space construction is basic to the intuition I am trying to capture with in the notion of 'category/sheaf - based process' or 'difference system'. In this Chapter, I will present a brief description of the Ehresmann structure, based on the expostion in my book "Geometry, Physics and Systems". For another exposotion, see the recent monograph "The Geometry of Jet Bundles", by D. J. Saunders.

2. The local maps.

All manifolds will be C^∞, finite-dimensional, and paracompact, unless mentioned otherwise. All maps between manifolds will also be "smooth," i.e., C^∞. Let X and Y be such manifolds.

Definitions. A local map from X to Y is a pair (U, ϕ) consisting of an open subset U of X and a C^∞ map $\phi: U \to Y$. Let **LM(X,Y)** be the space of all triples of the form

$$(x, U, \phi) \tag{2.1}$$

where:

$$U \text{ is an open subset of X} \quad , \tag{2.2}$$

$$\phi: U \to Y \text{ is a map} \quad , \tag{2.3}$$

$$x \in U. \tag{2.4}$$

$$\mathbf{LM_{sm}}(X,Y) = \text{the subset of } \mathbf{LM}(X,Y) \tag{2.5}$$
$$\text{consisting of the triples } (x, U, \phi)$$

such that:

$$\phi \text{ is a smooth, i.e. infinitely differentiable, map.} \qquad (2.6)$$

Remark. For physical intuition, think of 'x' as the 'label' of a 'particle' in a flow of matter in the space Y.

3. The source and target maps. Lagrangian and Eulerian description of material flows.

Construct maps:

$$\psi: \mathbf{LM}(X, Y) \to X \qquad (3.1)$$

$$\tau: \mathbf{LM}\,(X, Y) \to Y \qquad (3.2)$$

as follows:

$$\psi(x, \phi, U) = x \qquad , \qquad (3.3)$$

$$\tau(x, \phi, U) = \phi(x) \qquad . \qquad (3.4)$$

ψ is called the **source map**, and τ the **target map**.

Remark. These maps are the basic geometric objects underlying, respectively, the 'Lagrangian' and 'Eulerian' description of material flows in coninuum and fluis mechanics. Thus, x represents the 'Lagrangian' label of the particle while '$\phi(x)$' represnts the 'Eulerian position' of the particle.

4. The Ehresmann jet bundles.

We will construct the spaces of the Ehresmann theory as quotients of $\mathbf{LM}_{sm}\,(X, Y)$ under equivalence relations. The source and target maps σ and τ will be constant on these equivalence classes, and hence will pass to the quotient to define maps from the space of equivalence classes to X and Y, respectively. For the sake of notational simplicity, we will use the same letter to denote these maps on the equivalence classes.

First, consider the following equivalence relation:

$$(x, \phi, U) \sim (x', \phi', U') \tag{4.1}$$

iff:

$$x = x', \phi = \phi' \tag{4.2}$$

on an open subset of $U \cap U'$ containing x.

The quotient of $\mathbf{LM}(X, Y)$ under the equivalence relation (2.3) (i.e., the space of all equivalence classes) is called the **sheaf** of all maps: $X \to Y$, denoted as:

$$\mathbf{SM}(X, Y). \tag{4.3}$$

For each integer r, we define an equivalence relation as follows:

$$(x, \phi, U) \sim (x', \phi', U') \tag{4.4}$$

iff $x = x'$, ϕ and ϕ' **meet to order r at x**, in the sense that, in a local coordinate system for X about x, the partial derivatives at x of ϕ and ϕ' agree to order r.

The quotient of $\mathbf{LM}_{sm}(X, Y)$ by this equivalence relation is called the space of **rth-order jets** of $X \to Y$ denoted as:

$$J^r(X, Y) \tag{4.5}$$

It may be proved that the:

$$J^r(X, Y), r = 0, 1, 2, \dots , \tag{4.6}$$

are C^∞, finite-dimensional paracompact manifolds. The source and target maps:

$$\psi: J^r(X, Y) \to X \quad , \tag{4.7}$$

$$\tau: J^r(X, Y) \to Y \quad , \tag{4.8}$$

are submersions (i.e., the induced linear mapping on tangent bundles is onto), and hence define the Ehresmann jet spaces as fiber spaces.

Consider the following relation:

$$(x, \phi, U) \sim (x', \phi', U') \tag{4.9}$$

iff $x = x'$, ϕ and ϕ' meet to **infinite order** at x, in the sense that the Taylor expansions about x of ϕ and ϕ' in a local coordinate system agree.

The quotient of $\mathbf{LM}_{sm}(X, Y)$ by this equivalence relation is denoted by $J^\infty(X, Y)$, and called the space of **infinite-order jets**.

There are evident inclusions among the equivalence classes of these equivalence relations. (For example, two maps ϕ and ϕ' which agree to rth order at x, agree to order s, where s < r.) They lead to natural maps between the various quotient spaces:

$$\mathbf{SM}_{sm}(X, Y) \to J^\infty(X, Y)$$

$$\ldots \to J^r(X, Y) \to J^{r-1}(X, Y) \to \ldots \to J^0(X, Y) = X \times Y \tag{4.10}$$

Using the source and target maps, respectively, we obtain fiber space maps

$$J^r(X, Y) \to X \tag{4.11}$$

$$J^r(X, Y) \to Y \quad . \tag{4.12}$$

Differential forms on X and Y can be lifted up to $J^r(X, Y)$ via the pullback of the maps (4.11-4.12). For simplicity, we will often make no notational distinction between differential forms on X and Y and their pullback to $J^r(X, Y)$.

5. r-jet Prolongations of maps.

If:

$$\phi: U \to Y \tag{5.1}$$

is a local map from X to Y, its **r-jet** or **prolongation** is the map:

$$j^r(\phi): U \to J^r(X, Y) \tag{5.2}$$

obtained as follows:

> For $x \in U$, construct the element (x, ϕ) of $\mathbf{M}(X, Y)$ and (5.3)
> then assign it the equivalence class on $J^r(X, Y)$ to which it belongs.
> This is defined as: $j^r(\phi)(x)$. As x varies over U,
> the map (5.2) is obtained.

6. Prolongation of moving frames of differential forms.

There is a natural construction which leads from a pair of bases of differential forms on X and Y to basis of differential forms on $J^r(X, Y)$. We will now describe it after a brief recapitulation of the algebraic properties of vector fields on differential forms on a smooth manifold Z.

> $\mathbf{F}(Z)$ = ring of C^∞, real-valued functions (6.1)
> on the manifold Z.

For each integer $j \geq 0$, let:

$$\mathbf{D}^j(Z) = \text{the } C^\infty \text{ differential forms of degree j.} \tag{6.2}$$

$\mathbf{D}^j(Z)$ is an $\mathbf{F}(Z)$-module. The Exterior Multiplication operation

$$\wedge: \mathbf{D}^r(Z) \times \mathbf{D}^k(Z) \to \mathbf{D}^{j+k}(Z) \tag{6.3}$$

makes the collection

$$\mathbf{D}(Z) = \{\mathbf{D}^j(Z): \; j = 0, 1, 2, \ldots\} \tag{6.4}$$

into a graded, associative algebra. The exterior differentiation operation:

$$d: \mathbf{D}^j(Z) \to \mathbf{D}^{j+1}(Z) \tag{6.5}$$

is defined, with the usual properties.

A one-differential form θ on $J^r(X, Y)$ is a **contact form** if:

$$(j^r \phi) * (\theta) = 0 \tag{6.6}$$

for **each** local map $\phi: U \to Y$.

Let:

$$\mathbf{C}^r(X, Y) = \text{set of contact forms.} \tag{6.7}$$

It is an $\mathbf{F}(J^r(X, Y))$-module

Suppose:

$$\dim X = n \quad , \tag{6.8}$$

$$\dim Y = m \quad . \tag{6.9}$$

Choose the following ranges of indices and summation convention:

$$1 \leq i, j \leq n \quad , \quad 1 \leq a, b \leq m \quad . \tag{6.10}$$

Suppose that:

$$(\theta^a) \text{ is a basis for the one-forms } \mathbf{D}^1(Y) \tag{6.11}$$

$$(\text{"basis" as an } \mathbf{F} \ (Y)\text{-module})$$

and:

$$(\omega^i) \text{ is a basis for } \mathbf{D}^1 (X). \tag{6.12}$$

Consider these as 1-forms on $J^r(X, Y)$ pulled back via the projection map. Then, there are functions:

$$\left(y_i^a \ , \ y_{i_1 i_2}^a \ , \ \cdots \ , \ y_{i_1 \ldots i_r}^a \right) \tag{6.13}$$

on $J^r (X, Y)$ such that the forms

$$\eta^a = \theta^a - y_i^a \, \omega^i \quad , \tag{6.14}$$

$$\eta_{i_1}^a = d y_{i_1}^a - y_{i_1 i_2}^a \, \omega^{i_2} \quad , \tag{6.15}$$

$$\vdots$$

$$\eta^a_{i_1 \ldots i_{r-1}} = dy^a_{i_1 \ldots i_{r-1}} - y^a_{i_1 \ldots i_r} \, \omega^{i_r}$$

are a basis for $C^r(X, Y)$. Also, the forms:

$$\omega^i, \theta^a, dy^a_i, \ldots, dy^a_{i_1 \ldots i_r} \qquad\qquad (6.16)$$

form a basis for 1-forms on $J^r(X, Y)$.

Assume that the "moving frames" (ω^i), (θ^a) are differentials of coordinate systems, i.e. let us suppose that X and Y have a coordinate system, labeled:

$$x^i, y^a, \qquad\qquad (6.17)$$

with:

$$\omega^i = dx^i \quad , \qquad\qquad (6.18)$$

$$\theta^a = dy^a \quad . \qquad\qquad (6.19)$$

One then proves readily that the functions

$$\left(x^i, y^a, y^a_i, \ldots, y^a_{i_1 \ldots i_r} \right) \qquad\qquad (6.20)$$

associated with these bases form a coordinate system for:

$$J^r(X, Y) \quad .$$

A symbolic notation such as

$$(x, y, \partial_x y) \quad , \quad (x, y, \partial_{xx} y, \ldots) \qquad\qquad (6.21)$$

is often useful.

7   DIFFERENTIAL EQUATIONS, OPERATORS, AND SYMBOLS

Let X, Y, $J^r(X, Y)$, $r = 0, 1, 2, \ldots$, be as before. Work with a fixed coordinate system (x^i) and (y^a) for X, Y, and the associated coordinates $(x^i, y^a, y^a_i, \ldots)$ for $J^r(X, Y)$.

Definition: Let V be a real vector space, assumed finite-dimensional. An **rth-order differential operator symbol** with values in V is a (C^∞) map:

$$\sigma: J^r(X, Y) \to V \quad . \tag{7.1}$$

Given such a symbol, we will define a map

$$D_\sigma: LM(X, Y) \to LM(X, V) \quad , \tag{7.2}$$

which is the associated **differential operator**. For a map:

$$\phi: U \to Y,$$

U an open subset of X, $x \in U$, D_σ is defined by:

$$D_\sigma (\phi) (x) = \sigma (j^r (\phi) (x)) \quad . \tag{7.3}$$

The essence of this "geometric" approach to the theory of partial differential equations is the distinction between the "symbol," as a "**geometric object**," i.e., as a C^∞ mapping on a finite-dimensional manifold sitting above the domain and range manifolds X and Y ("independent" and "dependent" variables), and as a **mapping** between the **infinite-**dimensional space of local mappings. This correspondence is also implicit (and, in Lie's work, often quite explicit) in the 19th century literature on partial differential equations.

Definition: The **differential equation** associated with such a differential operator is the subset

$$DE \equiv \sigma^{-1}(0) \qquad\qquad (7.4)$$

of $J^r(X, Y)$. The **solutions** of the differential equations are the $\phi \in M(X, Y)$ such that

$$D_\sigma(\phi) = 0 \quad , \qquad\qquad (7.5)$$

or, equivalently,

$$j^r(\phi)(U) \subset DE(\sigma) \quad . \qquad\qquad (7.6)$$

Remark: This geometric way of defining the **geometric concept** of "differential equation" can be defined in many different "categories," e.g., the category of C^∞ maps, or "algebraic" maps, or, even with a little ingenuity, over "characteristic **p**." For the purpose of differential geometry and physics, the most important categories seem, at the present time at least, to be the C^∞ and the real or complex analytic maps.

8. PROLONGATIONS OF DIFFERENTIAL OPERATORS AND DIFFERENTIAL EQUATIONS

In the classical literature on nonlinear partial differential equations the process of **prolongation** is central, although not precisely defined. In Goldschmidt's and Spencer's work [9, 14, 15] one finds that it is an operator taking objects that live on a jet space of given order to one of next higher order. In this section I will describe a variant of their ideas.

With the notation explained in Section 7, let $\sigma: J^r(X, Y) \to V$ be a symbol map. A local mapping $\phi: X \to Y$, expressed in local coordinates

$(x, y(x))$ for X, Y, is then a **solution** of the associated differential equation if:

$$\sigma(x, y(x), \partial y(x)) = 0 \qquad \text{for all } x \in U \quad . \tag{8.1}$$

Now, for fixed i, $1 \le i \le n$, differentiate (8.1) with respect to x^i. An $(r + 1)$th-order differential equation is obtained for which ϕ is a solution. The symbol of these differential equations, which will be a map denoted as:

$$\delta_i \sigma: \ J^{r+1} (X, Y) \to V \quad , \tag{8.2}$$

will be a map on $J^{r+1} (X, Y)$ taking values in the vector space V.

For example, if $r = 1$, and if σ is a function of the coordinates

$$(x^i, y^a, y_i^a) \tag{8.3}$$

on $J^1(X, Y)$, then:

$$\delta_i \sigma \, (x^i, y^a, y_i^a, y_{ij}^a) \ = \ \frac{\partial \sigma}{\partial x^i} + \frac{\partial \sigma}{\partial y^a} \, y_i^a + \frac{\partial \sigma}{\partial y_j^a} \, y_{ij}^a \quad . \tag{8.4}$$

Remark: Notice in formula (8.4) that the "affine" structure of the jet spaces described and utilized by Goldschmidt [15] makes its appearance. As described partially in Ref. [16], it also plays a role in analytical mechanics.

We can now describe this in a more algebraic way by introducing the vector fields $V(X)$ on X. Suppose

$$V = A^i \, (x) \, \frac{\partial}{\partial x^i} \tag{8.5}$$

is such a vector field. Then, set:

$$\delta_V(\sigma) = A^i \, \delta_i \, (\sigma) \quad . \tag{8.6}$$

Let us sum up as follows:

Theorem 8.1: For each integer $r \geq 0$, let $\mathbf{F}\,(J^r\,(X,\,Y),\,V)$ be the space of C^∞ maps: $J^r\,(X,\,Y) \to V$. $\left[\text{Algebraically, } \mathbf{F}\,(J^r\,(X,\,Y),\,V) \text{ is } \mathbf{F}\,(J^r\,(X,\,Y)) \otimes V.\right]$ Thus, the prolongation described above is an $\mathbf{R}$-linear map:

$$\delta_V\colon \mathbf{F}\,(J^r\,(X,\,Y),\,V) \to \mathbf{F}\,(J^{r+1}\,(X,\,Y),\,V) \quad . \tag{8.7}$$

δ_V is a generalization of the usual Lie derivative.
The following formula holds:

For each map $\gamma\colon X \to Y$
$$\delta_V(j^r\,(\gamma)^*\,(\sigma)) = j^{r+1}\,(\gamma)^*\,(L_V\,(\sigma)) \quad . \tag{8.8}$$

Let $t \to \exp\,(tV)$ be the one-parameter pseudogroup on X generated by the vector field V. Then, for $x \in X$,

$$\delta_V(j^r\,(\gamma)^*\,(\sigma))\,(x) = \frac{\partial}{\partial t}\,(j^r\,(\gamma)^*\,(\sigma))\,[\exp\,(-\,tV)\,(x)]\Big|_{t\,=\,0} \quad . \tag{8.9}$$

9. PROLONGATION OF DIFFERENTIAL EQUATIONS

Continue with the notation of Section 8. Let V be a vector space, and

$$\sigma\colon J^r\,(X,\,Y) \to V$$

be an rth-order symbol map. The subspace:

$$\sigma^{-1}\,(0) \equiv DE\,(\sigma) \tag{9.1}$$

of $J^r\,(X,\,Y)$ defines a **differential equation**. A map $\phi\colon X \to Y$ is a **solution** if the following condition is satisfied.

$$j^r (\phi) (X) \subset DE (\sigma) \quad . \tag{9.2}$$

An alternate way of putting this is that:

$$j^r (\phi)^* (\sigma) = 0 \quad . \tag{9.3}$$

For each integer s > 0, and $V_1, \ldots , V_s \in \mathbf{V}(X)$, consider the generalized Lie derivative:

$$\mathbf{L}_{v_1} \ldots \mathbf{L}_{v_s} (\sigma) \quad . \tag{9.4}$$

It is a map $J^{r+s} (X, Y) \to W$. For each integer m, $m \geq s$, we can pull back the functions (9.4) to live on $J^{r+m} (X, Y)$, via the "forgetting" map: $J^{r+m} \to J^{r+s}$. In this way, we get a collection:

$$\left\{ \sigma, \mathbf{L}_{v_1} (\sigma), \ldots , \mathbf{L}_{v_1} \cdots \mathbf{L}_{v_m} (\sigma) \colon V_1, \ldots , V_m \in \mathbf{V} (X) \right\} \tag{9.5}$$

of maps $J^{r+m} (X, Y) \to W$. Let $DE^m (\sigma)$ be the subset of points of $J^{r+m} (X, Y)$ on which **all** the maps (9.5) **vanish**. Of course, as a differential equation, $DE^m(\sigma)$ has precisely the same set of solutions as $DE (\sigma)$. However, it is the geometric properties of each $DE^m (\sigma)$ that are of greatest interest.

In particular, one is interested in knowing something of the algebro-geometric and topological properties of $DE^m (\sigma)$ as subset of manifolds. Similar questions about subsets of jet spaces defined by equations occur in the theory of singularity of mappings. [17] In this discipline, work can be done effectively in the category of C^∞ mapping; however, at the partial differential equation level, in the current state of mathematical development, serious work will require the assumption that the manifolds X, Y and the symbol map σ be **real analytic**. When discussing the notion of **general** and **singular** solutions the distinction C^∞ and real-analytic category becomes critical. In fact, it is well known in the classical theory of partial differential equations that there are major qualitative differences between the C^∞ and real analytic category. If the coefficients of these equations were assumed analytic, the classic Cauchy-Kowalewsky theorem would guarantee the existence at least of local solutions, while this general result breaks down in the C^∞ case.

10. GENERAL SOLUTIONS OF SYSTEMS OF REAL-ANALYTIC PARTIAL DIFFERENTIAL EQUATIONS

Continue with the set up described in previous sections: X, Y, $J^r (X, Y)$, $\sigma: J^r (X, Y) \to V$, $DE (\sigma) \subset J^r (X, Y)$, the mth prolongation $DE^m (\sigma) \subset J^{r+m} (X, Y)$, for $m \geq 1$. Now, assume that the data is **real analytic**. $DE^m (\sigma)$ is then a real analytic subvariety of the real analytic manifold $J^{r+m} (X, Y)$. It is then a union of a finite number of irreducible subvarieties.[18] We can then consider the analytic real-valued functions $f: U \to R$ defined on open subsets U of $J^{r+m} (X, Y)$, which are identically zero on at least one irreducible branch of $DE^m (\sigma)$. Denote these functions by $DE^m (\sigma, F)$. (They form a sheaf.) Let GS be a family of maps: $X \to Y$, such that each $\phi \in GS$ is a solution of $DE (\sigma)$,

$$j^r (\phi)^* (\sigma) = 0 \quad . \tag{10.1}$$

For each integer $m \geq 0$, each open subset $U \subset J^{r+m} (X, Y)$, we can consider the set of maps $f: U \to R$ such that

$$J^r (\phi)^* (f) = 0 \quad \text{for all } \phi \in GS \quad . \tag{10.2}$$

Definition: GS is a **general solution** of the differential equations associated with the symbol map $\sigma: J^r (X, Y) \to V$ if each function f satisfying (10.2), for an integer $m \geq 0$, belongs to $DE^m (\sigma, F)$.

In words, this condition means that each differential equation satisfied by every map ϕ in the family GS is contained in the differential equation obtained by prolonging σ by differentiation.

This definition of "general solution" is one of two given and used in the 19th century literature [1, 5]; it is due to Ampère. The other is given Darboux' name, but we will not go into it here.

These generalities are best comprehended by thinking about certain simple examples.

11. THE GENERAL SOLUTION OF THE ONE-DIMENSIONAL WAVE EQUATION

Here,

$$X = R^2 \quad , \quad \text{with coordinates } (x^1, x^2),$$

$$Y = R \quad , \quad \text{with coordinates } y.$$

$J^2 (X, Y)$ has coordinates:

$$z = (x, y, t_1, y_2, y_{11}, y_{12}, y_{22}).$$

Define σ: $J^2 (X, Y) \to R$ as follows:

$$\sigma (z) = y_{12} \quad . \tag{11.1}$$

The solutions of the differential equations are the maps

$$x \to y (z)$$

such that:

$$\frac{\partial y}{\partial x_1 \, \partial x_2} = 0 \quad . \tag{11.2}$$

This is (in "light cone" or "characteristic" coordinates) just the wave equations, in one space, one time variable.

To define the prolongations of (11.1), let $1 \leq i, j, i_1, \ldots, \leq 2$ be indices

$$(x, y, y_i, y_{ij}, \ldots)$$

are coordinates of $J^\infty (X, Y)$. The prolonged varieties are the subsets of $J^{r+m}(X, Y)$ obtained by setting equal to zero the functions:

$$y_{i \cdots i_s 12} \, , \tag{11.3}$$

where $1 \leq i_1, \ldots, i_s \leq 2$, s an arbitrary nonnegative integer.

The classical "general solution" of (7.3) is the family of maps $R^2 \to R$, of the form:

$$\phi (x_1, x_2) = g_1 (x_1) + g_2 (x_2) \, , \tag{11.4}$$

where g_1, g_2 are arbitrary maps: $R \to R$.

It is now obvious that any map f: $J^{r+m} \to R$ that satisfies (10.2) is a function **only** of the variables (11.3), which is the condition required to verify that the family of maps (11.4) is a general solution of the DE (11.2).

12. THE CLAIRAUT EQUATION: ITS "GENERAL" AND "SINGULAR" SOLUTION

The classical Clairaut equation is a first-order ordinary differential equation of the form

$$y(x) - x\frac{dy}{dx} - f\left(\frac{dy}{dx}\right) = 0 \tag{12.1}$$

to be solved for a map $x \to y(x)$ of $R \to R$. f is given C^∞ map: $R \to R$.

To put this into the standard framework described above, let:

$$X = Y = R \quad . \tag{12.2}$$

Denoting the coordinates of $J^1(X, Y)$ by (x, y, y'), the symbol σ is the map:

$J^1(X, Y) \to R$ defined as follows:

$$\sigma(x, y, y') = y - xy' - f(y') \quad . \tag{12.3}$$

In order to define the prolonged systems, let us differentiate (8.1):

$$0 = \frac{dy}{dx} - \frac{dy}{dx} - x\frac{d^2y}{dx} - f'\left(\frac{dy}{dx}\right)\frac{d^2y}{dx^2}$$

$$= -\frac{d^2y}{dx^2}\left(x + f'\left(\frac{dy}{dx}\right)\right) \quad . \tag{12.4}$$

Thus,

$$\delta\sigma((x, y, y', y'')) = y''(x + f'(y')) \quad . \tag{12.5}$$

The variety of $J^2(X, Y)$ obtained by setting:

$$0 = \sigma = \delta$$

is **reducible**.

At the differential equation level, C^∞ solutions of (12.4) are of two sorts: Solutions of

$$\frac{d^2 y}{dx^2} = 0 \tag{12.6}$$

or

$$f' \left(\frac{dy}{dx} \right) = - x \quad . \tag{12.7}$$

The "general solution" of (8.6) is:

$$y = ax + b \quad , \tag{12.8}$$

where a and b are real constants. The condition that it be a solution of (12.1)
introduces a relation between a and b:

$$ax + b - xa - f(a) = 0 \ ,$$

or

$$f(a) = b \quad . \tag{12.9}$$

Theorem 12.1: The family

$$\phi_a(x) = ax + f(a) \quad , \tag{12.10}$$

parametrized by $a \in R$, is a general solution, in the Ampère sense, of the differential equation (12.1).

Proof: Suppose:

$$a \left(x, y, y', \dots , y^{(r)} \right)$$

in an rth-order differential operator symbol, for which $y(x) = \phi_a(x)$ is a solution for each a. Now,

$$\frac{d\phi_a}{dx} = a \quad ,$$

$$\frac{d^2\phi_a}{dx^2} = 0 \quad ,$$

$$\cdot$$
$$\cdot$$
$$\cdot$$

Hence,

$$a \ (x, \ ax + f \ (a), \ a, \ \dots \ , \ 0) = 0 \quad , \tag{12.11}$$

and then

$$a \ (x, \ y, \ z, \ \dots \ , \ 0) = 0 \qquad \text{for all x, y} \quad . \tag{12.12}$$

We conclude that a must be for the form:

$$a = \beta_1 y'' + \dots \beta_{r-2} \ y^{(r)} \quad ,$$

where $\beta_1, \ \dots \ , \beta_{r-2}$ are arbitrary functions on $J^r \ (X, \ Y)$. This is what is required to prove the ϕ_a are "general solutions."

The Clairaut equation is the example traditionally used in the classical treatises to introduce the notion of "singular solution." They are the solutions of the original differential equation (12.1), which are not part of the general solution (12.10). They must then be solutions of (12.7). The general solution of (8.7) depends on one real parameter, say

$$x \rightarrow y^s \ (x; \ c) \tag{12.13}$$

with

$$f' \left(y_x^s \, (x; \, c) \right) = - x \quad . \tag{12.14}$$

Now, the condition that the family of functions of x, indicated in (12.13), **also** satisfy the original differential equation, is

$$y^s \, (x, \, c) - xy_x^s - f \, (y_x^s) = 0 \quad . \tag{12.15}$$

In general, (12.14) and (12.15) will have solutions for only a discrete set of c's. These are the **singular solutions** of the Clairaut differential equation (12.1).

Now, the traditional geometric interpretation of these singular solutions of the Clairaut equation is that they are **the envelopes** of the family of curves in R^2 given by the general solution

$$x \to (x, \, y \, (x; \, a)) = (x, \, ax + f \, (a)) \quad . \tag{12.16}$$

For each a, (12.16) is a straight line. They are the tangent lines to the curves $x \to (x, \, y^s \, (x))$, where y^s are the solutions of (12.14) and (12.15).

13. THE DIFFERENTIAL ALGEBRA OF DIFFERENTIAL FORMS ON THE INFINITE ORDER JET SPACE

Now return to the usual situation where X and Y are real, C^∞, finite-dimensional manifolds. We have defined the space $J^\infty \, (X, \, Y)$ of **infinite order jets** of mappings: $X \to Y$. We have the tower of maps

$$J^\infty \, (X, \, Y) \to J^{r+1} \, (X, \, Y) \to J^r \, (X, \, Y) \to \dots \quad . \tag{13.1}$$

$J^\infty \, (X, \, Y)$ is not a manifold in the usual sense; however, the tower (13.1) can be used to attach a **differential form algebra** to it.

For each integer $r \geq 0$, let

$$\mathbf{D}\,(J^r\,(X,\,Y)) = \left\{\,\mathbf{D}^n\,(J^r(X,\,Y)):\ n = 0,\,1,\,\dots\,;\,\wedge\,\right\} \qquad (13.2)$$

be the graded associative differential algebra of differential forms on the manifold $J^r\,(X,\,Y)$. The dual of maps on forms to the tower (13.1) leads to a set of **inclusion** relations

$$\mathbf{D}\,(J^0(X,\,Y)) \subset \mathbf{D}\,(J^1(X,\,Y)) \subset \dots \qquad (13.3)$$

among the differential forms on the tower of jet spaces. We can then define a differential form algebra as the **union** of this sequence of sets. We denote it as $\mathbf{D}\,(J^\infty(X,\,Y))$, and call it the **differential form algebra** of the infinite order jet space. Thus, a differential form of degree n on the $J^\infty(X,\,Y)$ is, by definition, an element of $\mathbf{D}^n\,(J^r\,(X,\,Y))$ for r sufficiently large. The operations d, $\wedge$ on these forms are just those inherited from the corresponding operations on the finite-order jet manifolds, where they are defined by the usual calculus-on-manifold apparatus.

However, $\mathbf{D}\,(J^\infty(X,\,Y))$ has an additional algebraic structure that it inherits from the **prolongation** operation on functions defined in Section IV. Thus, to every vector field V on X we have assigned an operation:

$$\delta(V)\colon\ \mathbf{F}(J^r\,(X,\,Y)) \rightarrow \mathbf{F}(J^{r+1}(X,\,Y)) \qquad . \qquad (13.4)$$

$\delta(V)$ is a **derivation** with respect to the algebraic structure on $\mathbf{F}$ $(J^r(X,\,Y))$:

$$\delta(V)\,(f_1 f_2) = \delta\,(V)(f_1)\,f_2 + f_1\delta(V)(f_2) \qquad \text{for}$$
$$f_1,\,f_2 \in \mathbf{F}(J^1(X,Y)) \qquad . \qquad (13.5)$$

There is a convention inherent in the right-hand side of (13.5). $\delta\,(V)(f_1)$ is an element of $\mathbf{F}\,(J^{r+1}(X,Y))$, whereas f_2 is an element of

$F(J^r(X,Y))$. In order to multiply them, $F(J^r(X, Y))$ must be identified with a subset $\big[$as it is via the dual of the projection map $J^{r+1}(X, Y) \to J^r(X, Y)$

of $F(J^{r+1}(X, Y))\big]$.

Now, $F(J^{r+1}(X, Y))$ is $D^0(J^r(X, Y))$, and $D^0(J^\infty(X, Y))$ is, by definition, the union of all the $D^0(J^r(X, Y))$, for $r = 0, 1, \dots$. $\delta(V)(f)$ is the same whether it comes from $J^r, J^{r+1}, \dots$ (of course, this is inherent in the geometric meaning of $\delta(V)$!). Hence, $\delta(V)$ can be defined in a uniform way as a **derivation** of the algebra:

$$F(J^\infty(X, Y)) \equiv D^0(J^\infty(X, Y)) \quad ,$$

such that:

$$\delta(V)\big(D^0(J^r(X, Y))\big) \subset D^0(J^{r+1}(X, Y)) \quad \text{for } r = 0, 1, \dots \quad .(13.6)$$

We can now extend $\delta(V)$ to act on higher degree differential forms, in a way which commutes with exterior derivation and acts as a derivation on exterior multiplication. Suppose, for example, that (x) denotes coordinates on X, (y) coordinates on Y:

$$(x, y, \partial_x y, \partial_{xx} y, \dots)$$

denote coordinates on $J^\infty(X, Y)$. A differential form on $J^\infty(X, Y)$ can be written as follows::

$$\omega = a\,df_1 \wedge \dots df_m + \dots \quad ,$$

with $a, f_1, \dots, f_m \in D^0(J^\infty(X, Y))$. Thus,

$$\delta(V)(\omega) = \delta(V)(a)\,df_1 \wedge \dots \wedge df_m$$
$$+ a d\big[\delta(V)(f_1)\big] \wedge \dots \wedge df_m + \dots \quad . \qquad (13.7)$$

We have:

$$d\delta(V) = \delta(V)d \qquad \text{for } V \in \mathbf{V}(X) \quad . \qquad\qquad (13.8)$$

Thus, we can translate "geometry" to "algebra" by assigning to (X, Y) the associative differential form algebra:

$$\mathbf{D}(J^{\infty}(X, Y))$$

with its graded, differential structure. (This is, in fact, an excellent example of a functor assocted with two categories.) Many of the jewels which lie hidden in the 19th century theory of differential equations and the calculus of variations will be found when this algebraic structure is completely analyzed and understood.

Remark: A special case of this construction was given in Chapter 10 of Interdisciplinary Mathematics, Vol. XII. [13] Here, I have only touched on an area that is extensively developed in the work of Kupershmidt and Vinogradov. [22, 23]

1   E. Goursat, **Lecons sur l'intégration des équations aux derivés partielles du second ordre** (Gauthier-Villars, Paris, 1890), Vols. I and II.

2   G. Darboux, **Lecons sur la théorie générale des surfaces** (Gauthier-Villars, Paris, 1894, 1915).

3   E. von Weber, G. Floquet, and E. Goursat, "Proprietés générales des systems d'equations aux derivées partielle," in **Encyclopédie des Sciences Mathématiques** (Gauthier-Villars, Paris, 1913), Tome II, Vol. 4.

4   C. Riquier, **Les systémes d'equations aux derivées partielles** (Gauthier-Villars, Paris, 1910).

5 A. Forsythe, **Theory of Differential Equations** (Dover, New York, 1959).

6 E. Picard, **Traité d'analyse** (Gauthier-Villars, Paris, 1922).

7 C. Ehresmann: (a) "Sur les structures infinitésimales régularières," Congrés Int. Math. Amsterdam **1**, 479–80 (1954); (b) "Connexions infinitésimales," Colloq. Top. Alg. Bruxelles, 29–55 (1950); (c) "Structures infinitésimales et pseudogroupes de Lie," Colloq. Int. C.N.R.S. Géom. Diff. Strasbourg, 97–110 (1953); (d) C. R. Acad. Sci. Paris **240, 246**, 360 (1958) (1954); **241**, 397, 1755 (1955); (e) "Connexions d'ordre supérieur," **Atti 5 Cong. dell'Unione Mat. Italiana, 1955** (Cremonese, Rome, 1956), pp. 326–8; (f) "Categories topologiques et catégories différentiables," Colloq. Géom. Diff. Globale Bruxelles, C.B.R.M., 137–150 (1958); (g) "Groupoides differenciales," Rev. Un. Mat. Argentina Buenos-Aires **XIX**, 48 (1960).

8 R. Hermann, "Some differential geometric aspects of the Lagrange variational problem," Ill. J. Math. **6**, 634–73 (1962).

9 A. Kumpera and D. C. Spencer, **Lie Equations**, Ann. of Math. Studies **73** (Princeton U. P., Princeton, NJ, 1972).

10 R. Hermann, **Geometry Physics and Systems** (Marcel Dekker, New York, 1973).

11 E. Ince, **Ordinary Differential Equations** (Dover, New York, 1956).

12 E. T. Whittaker, **A Treatise on Analytical Dynamics of Particles and Rigid Bodies** (Cambridge U. P., London and New York, 1959).

13 R. Hermann, **Geometry of Non-Linear Differential Equations, Bäcklund Transformations and Solitons, Part**

A, Vol. XII of Interdisciplinary Mathematics (Math. Sci. Press, Brookline, MA, 1976).

14 D. C. Spencer, "Overdetermined systems of linear partial differential equations," Bull. Am. Math. Soc. **75**, 179–239 (1969).

15 H. Goldschmidt, (a) "Nonlinear partial differential equations," J. Diff. Geom. **1**, 269–307 (1967); (b) "Sur la structure des équations de Lie," J. Diff. Geom. **6**, 357–73 (1972); **7**, 67–95 (1972).

16 R. Hermann, "The differential geometry of general mechanical systems from the Lagrangian point of view," J. Math. Phys. **23**, 2077 (1982).

17 M. Golubitsky and V. Guillemin, **Stable Mappings and Their Singularities** (Springer-Verlag, New York, 1973).

18 H. Whitney, **Complex Analytic Varieties** (Addison-Wesley, Reading, MA, 1972).

19 E. Vessiot, "Sur une théorie nouvelle des problèmes généraux d'integration," Bull. Soc. Math. Fr. **52**, 336–96 (1924).

20 E. Cartan, **Les systèmes éxterieures et leurs applications géométriques** (Herman, Paris, 1946).

21 R. Hermann, "E. Cartan's geometric theory of partial differential equations," Adv. in Math. **1**, 265–317 (1965).

22 B. A. Kupershmidt, in **Lecture Notes in Mathematics 775** (Springer-Verlag, New York, 1980), pp. 162–218.

23 A. M. Vinogradov, "Many-valued solutions, and a principle for the classification of nonlinear differential equations," Dokl. Akad.

Nauk **SSSR 210**, 11–14 (1973) [Sov. Math. Dokl. **14**, 661–5 (1973)], MR 50 No. 1294.

24 R. Hermann, "The second variation for variational problems in canonical form," Bull. Am. Math. Soc. **71**, 145–8 (1965).

25 R. Hermann, "The second variation for minimal submanifolds," J. Math. Mech. **16**, 473–92 (1966).

26 R. Hermann, **Differential Geometry and the Calculus of Variations** (Math. Sci. Press, Brookline, MA, 1977), 2nd ed.

REMARKS ON DIFFERENTIAL SYSTEMS

1. THE GEOMETRIC FORMULATION OF THE DIFFERENTIAL SYSTEMS CONCEPT

The mathematical technology of "**differential equations**" is of course at the heart of the applicability of mathematics to much of the physical world. My goal in this work is to carry over some of this power to computer science, replacing "differential" by "difference." However, what will be most important to us will not be the traditional **analytic** idea of differential equations, but the **geometric** and **algebraic**.

In contemporary differential geometry, there are two complementary approaches to the theory of differential systems. One proceeds via the Ehresmann— Spencer-Goldschmidt theory of jet spaces , the other via E. Cartan's theory of exterior differential systems . In this paper, I will emphasize the second approach.

Let us begin with the traditional, coordinate-dependent point of view toward the theory of systems of partial differential equations. Choose notations as follows. Let

$$y = (y^1, \dots , y^m) \in R^m$$

$$x = (x^1, \dots , x^n) \in R^n$$

be a system of **dependent** and **independent** variables, respectively. Let

$$y_x, y_{xx}, \dots$$

be auxiliary variables which represent the first, second, ... , partial derivatives of the y as a function of the x. Thus,

$$y_x \in R^{nm}$$

$$y_{xx} \in R^{mn^2}$$

and so on.

Definition. A **system of partial differential equations of order** r is determined by a map

$$f: \ R^m \times R^n \times R^{nm} \times R^{nm^2} \times \ldots \times R^{nm^r} \to R^p \tag{1.1}$$

$$(x, \ y, \ y_x, \ y_{xx}, \ \ldots) \to f\,(x, \ y, \ y_x, \ y_{xx}, \ \ldots) \quad .$$

A **solution** of this system is a map

$$R^n \to R^m$$
$$x \to y\,(x)$$

such that:

$$f\left(x, \ y\,(x), \ \frac{\partial y}{\partial x}, \ \frac{\partial^2 y}{\partial x^r} \ \ldots \ \frac{\partial^2 y}{\partial xx^r}\right) = 0 \quad . \tag{1.2}$$

The exterior systems approach towards such a system proceeds in the following way. Construct a manifold Z of variables

$$x, \ y, \ y_x, \ y_{xx}, \ \ldots \quad . \tag{1.3}$$

Introduce the exterior differential system C (i.e., the differential ideal in the differential algebra of differential forms on Z) generated by the following 1-forms on Z:

$$dy - y_x \, dx \tag{1.4}$$
$$dy_x - y_{xx} \, dx \tag{1.5}$$
$$\vdots$$

C is called the **contact system**. Now, let **E** be the exterior system on Z generated by C and the 0-forms that make up the components of the map f that appears on the right-hand side of (1.2).

Let us now formulate standard results in the following way:

Theorem 1.1. Let $X = \{x\}$ be the domain manifold, $Y = \{y\}$, the range manifold, for the system (1.3) of partial differential equations. Let

$$Z = \{x, y, y_x, y_{xx}, \dots\} \tag{1.6}$$

be the space of the indicated variables (in terms of the theory of jets, $Z = J^r(X, Y)$ where the integer r is the order of the system of differential equations (1.2)). Let

$$\alpha: Z \to X \tag{1.7}$$

$$\beta: Z \to Y \tag{1.8}$$

be the maps defined as follows:

$$\alpha(x, y, y_x, y_{xx}, \dots) = x \tag{1.9}$$

$$\beta(x, y, y_x, y_{xx}, \dots) = y \quad . \tag{1.10}$$

α is called the **source map**; β the **target map**.

Let **E** be the exterior differential system on Z generated by the contact forms C and the system functions occurring on the left-hand side of (1.2). Then the (smooth) solution maps

$$\phi: x \to y(x)$$

$$X \to Y \tag{1.11}$$

of the system (1.2) are identical to the maps of the form

$$\phi = \beta^{\gamma} \quad , \tag{1.12}$$

where $\gamma: X \to Z$ is a (smooth) map satisfying the following conditions:

$$\gamma^* (\mathbf{E}) = 0 \quad , \tag{1.13}$$

i.e., γ is an **integral submanifold** of the exterior system $\mathbf{E}$, and

$$\alpha\gamma = \text{identity} \quad . \tag{1.14}$$

This statement of standard results enables us now to give a general formulation of what is meant by a "differential system."

Definition. Let X and Y be finite dimensional, paracompact C^{∞} manifolds. A **differential system** on (X, Y) is defined by attaching the following geometric structures to (X, Y):

a) A manifold Z, together with smooth maps
$$\alpha: Z \to X$$

$$\tag{1.15}$$

$$\beta: Z \to Y \quad .$$

b) An exterior differential system $\mathbf{E}$ on Z.

The whole system is denoted as follows:

$$(\alpha, \beta, \mathbf{E}) \quad . \tag{1.16}$$

Of course, "equations" of any type involve two sorts of objects: The **solutions** and the relations determining them.

Definition. Let $\alpha: Z \to X$, $\beta: Z \to Y$, $\mathbf{E}$, determine a differential system. A map

$$\phi: X \to Y$$

is called a **solution** of the system if there is a smooth map

$$\gamma\colon X \to Z$$

satisfying the following conditions:

$$\gamma^*(\mathbf{E}) = 0 \quad , \tag{1.17}$$

i.e., γ is an integral map for the exterior system $\mathbf{E}$.

$$\alpha\gamma = \text{identity} \quad \text{map} \tag{1.18}$$

$$\beta\gamma = \phi \quad . \tag{1.19}$$

Having identified a space of "solutions," the next step is the study of **morphisms**.

2. MORPHISMS OF DIFFERENTIAL SYSTEMS

Having defined the **objects** called "differential systems," the next step should be to define the "arrows" or **morphisms**. In fact, there are several natural candidates in the literature, and each has an important physical or mathematical meaning and background.

The simplest situation occurs if one deals purely with an exterior system.

Definition. Let $\mathbf{E}$ and $\mathbf{E}'$ be exterior differential systems on manifolds Z and Z'. A map

$$\phi\colon Z \to Z'$$

is said to be a **morphism** between them if the following condition is satisfied:

$$\phi^*(\mathbf{E}') \subset \mathbf{E} \quad . \tag{2.1}$$

Suppose now that

$$\alpha\colon Z \to X, \ \beta\colon Z \to Y$$

$$\mathbf{E} \subset \mathbf{D}\,(Z) \quad , \tag{2.2}$$

$$\alpha': \ Z' \to X', \beta': \ Z' \to Y' \tag{2.3}$$

$$\mathbf{E'} \subset \mathbf{D}\,(Z')$$

are two differential systems, defining solutions as maps $X \to Y$ and $X' \to Y'$. The logically simplest definition to define "morphisms" is as follows:

Definition. A morphism, in the first sense, between systems (2.2) and (2.3), is a pair of maps

$$\phi_1: \ X' \to X$$

$$\phi_2: \ Y \to Y' \tag{2.4}$$

satisfying the following condition:

If $\gamma: \ X \to Y$ is a solution of system (2.2), then the map

$$\gamma' = \phi_2\,\gamma\,\phi_1: \ X' \to X' \tag{2.5}$$

is a solution of the system (2.3).

This is the most straightforward notion. Here is a more general version.

Definition. Associate to each solution map

$$\phi: \ X \to Y$$

of the system its **graph**:

$$gr\,(\phi): \ X \to X \times X$$
$$x \to (x, \, \phi\,(x)) \quad .$$

Then, a map

$$\gamma\colon X \times Y \to X' \times Y'$$

is a **morphism of the second type** from system (2.2) to system (2.3) if the following condition is satisfied:

For each solution map $\phi\colon X \to Y$ of (2.2), there is a solution map $\phi'\colon X' \to Y'$ of (2.2) such that:

$$\gamma\,(\text{graph }(\phi)\,(X)) \subset \text{graph }(\phi')\,(X) \qquad . \qquad (2.6)$$

Finally, we shall describe a third version:

Definition. A map $\delta\colon Z \to Z'$ is said to define a **morphism of the third kind** from system (2.2) to system (2.3) if the following conditions are satisfied:

a) δ is a morphism of the exterior system $\mathbf{E}$ to the exterior system $\mathbf{E'}$ in the sense that condition (2.1) is satisfied. Further,
$$X = X' \qquad . \qquad (2.7)$$

b) For each solution pair
$$\gamma\colon X \to Z$$
$$\phi\colon X \to Y \quad ,$$

of (2.2) such that:
$$\alpha\gamma = \text{identity} \qquad\qquad (2.8)$$
$$\phi = \beta\gamma \qquad . \qquad\qquad (2.9)$$

There is a similar solution pair of (2.3),

$$\gamma'\colon X' \to Z'$$
$$\phi'\colon X \to Y$$

such that:

$$\gamma' = \delta\gamma \qquad . \qquad\qquad (2.10)$$

Other versions of the basic idea are possible. This is also tied into the physicist's idea of the concept of **symmetry** of a system of differential equations.

3. SUBSYSTEMS OF DIFFERENTIAL SYSTEMS

Any reasonable definition of a "morphism" between two differential systems would, of course, contain within it the definition of **subsystem** as a special case. However the definition of "subsystem" is so important — for example, as we shall see in Section 5, it is closely linked to "feedback" for input-output control systems — that it is worth our while to give an independent treatment.

Definition. Let

$$\alpha\colon Z \to X, \beta\colon Z \to Y$$
$$\mathbf{E} \subset \mathbf{D}\,(z)$$

be data which determine a differential system for maps from X to Y. A **subsystem** of the system $(\alpha, \beta, \mathbf{E})$ is defined by an exterior differential system $\mathbf{E}'$ on Z such that:

$$\mathbf{E} \subset \mathbf{E}' \quad . \tag{3.1}$$

Here is the main geometric property:

Theorem 3.1. Let $(\alpha, \beta, \mathbf{E}')$ be a subsystem of $(\alpha, \beta, \mathbf{E})$, in the sense that (3.1) is satisfied. Then, each solution map

$$\phi\colon X \to Y$$

of system $(\alpha, \beta, \mathbf{E}')$ is also a solution of $(\alpha, \beta, \mathbf{E})$.

Proof. Follows from the definition and condition (3.1).

4. DISCRETIZATION OF DIFFERENTIAL SYSTEMS VIA THE THEORY OF EXTERIOR DIFFERENTIAL SYSTEMS

A major theme in applied mathematics is the replacement and approximation of systems of **differential** equations by **difference equations**. From the geometric point of view, it would be desirable (and useful) to be able to do this in a way that preserves as far as possible the functorial properties of differential systems relative to the category of smooth manifolds and smooth maps between them. Although I know of no precise and/or general theorems along these lines, it is clear (at least intuitively) that this is not possible in complete generality. (In fact, I believe that there are analogies between this mathematical problem and that of formulating a **covariant quantization** process that would assign a **quantum-mechanical system** to each **classical** mechanical system in such a way as to be (at least partially) "functorial" relative to the category of smooth manifolds and maps.) In this section, I present one algorithm for the discretization process based on Cartan's theory of exterior differential systems.

Let us consider an exterior system $\mathbf{E}$ on a manifold Y which is generated by 1-forms. We will only work locally, in terms of a coordinate system that we will label (vectorially) as follows:

$$y = (y^a) \quad , \quad 1 \leq a, b \leq m \quad .$$

Suppose that $\mathbf{E}$ is generated by 1-forms written vectorially as follows:

$$\theta = a \ (y) \ dy \tag{4.1}$$
$$y \in R^m \quad .$$

An **integral submanifold** is a map

$$\phi: x \rightarrow y \ (x)$$

such that:

$$\phi^* \ (\theta) = 0 \quad . \tag{4.2}$$

Condition (4.2) can be written as a system of PDE's as follows:

$$a\,(y\,(x))\,\frac{\partial y}{\partial x} = 0 \qquad . \tag{4.3}$$

We can now formulate the discretization process in the following geometric way. Replace the (vector-valued) differential forms dx and dy by Δx. (4.2) is to be replaced by a difference equation in the following form:

$$a\,\Delta y = 0 \qquad . \tag{4.4}$$

The expression on the left-hand side of (4.4) will be called an **exterior difference form**.

The methods used here may be readily extended to more complicated systems, a task we leave to a later point.

REMARKS ON COMPUTATION AND APPROXIMATION

1. INTRODUCTION

Control Theory and Artificial Intelligence have much in common. Both are essentially engineering disciplines, but with roots in mathematics. In the case of Control, these roots are in differential equations, matrix algebra, and differential geometry, while for AI they are in mathematical logic. However, the most significant common interests lie in the realm of goals.

Control Theory is concerned with interactions between machines -- with more-or-less known characteristics -- and the environment, or with designing rather idealized machines that will act on certain inputs of known mathematical characteristics to achieve certain outputs, perhaps with limited learning features ("identification" and "adaptation"). AI has not grown out of such relatively concrete and modest engineering situations as Control, and has taken on the vastly more ambitious program of simulating human thought and behavior in a non-trivial way (the "Turing Test"). As a result of this difference of emphasis, they have had almost nothing to do with one another.

Since its origins in the 1950's, Control has been closely linked to both the mathematical technology of differential equations and matrix algebra and to the development of computer hardware and software that can handle this sort of mathematics with facility. The period of its maximal impact on engineering was undoubtedly the 1960's, when there was a remarkable synergism and resonance between these two components. On the one hand, there were important advances on the mathematical side: the Kalman filter, linear quadratic and matrix Riccati methods, observer theory, pole placement and Brunovsky canonical form, the Extended Kalman Filter as a first approximation to nonlinear filtering, the development of methods of stochastic processes and the Ito-Stratonovich stochastic calculus, the beginning of adaptive control, the first applications of the Caratheodory - Chow theorem and Lie theoretic methods to nonlinear control, and so on. These mathematical methods were well-suited to the computers of the day:

Let us call it the era of FORTRAN and NUMERICAL CONTROL. This combination was enormously important and influential in the technology of the 1960's and beyond: The APOLLO program, air and spacecraft control, control of chemical processes, etc. The challenge that faces us today is to extend the ideas of 1960's style control theory to take advantage of the current computer technology, especially the advances in symbolic computation centered around the computer languages LISP and PROLOG.

Another way of thinking about the problem of developing a "symbolic" control theory is that we must create a computer substitute or equivalent for the description of the world provided to us by differential equations. This is one of the primary problems of the whole AI enterprise [10, 20, 22, 23]. Much of contemporary computational logic is also oriented towards this goal, and provides us with computational tools that should be useful in the engineering world as LISP [6] and symbolic computation [27] becomes more widespread. I believe that there is some sort of half-way house between what the AI people have done in the last twenty years -- which, in my opinion, still lacks that "hardness" and precision that is needed for serious engineering applications -- and the traditional differential-equations oriented description used in engineering and physics.

From the engineer's point of view, much of the task of re-orienting Control practice and theory in a "symbolic" direction involves learning how to most efficiently perform the traditional Control tasks in the context of LISP. Indeed, this is itself a system-theoretic problem, and it might even be of interest to approach it from that point of view. After all, LISP is itself a highly mathematical language -- one designed to compute functions defined recursively. Much of control engineering can be considered as a map from inputs to outputs, where the input and output spaces have a specified algebraic structure. Perhaps one should be looking for some sort of Realization Theory, where the analogue of the State Space is the LISP program itself? It would be very attractive if it turned out that there was some sort of "minimal" realization of this sort! Now, writing a computer program is, in logical terms, the same as giving the proof of a theorem, with certain very well codified rules used in the deduction. Thus, we must think about theorem proving from a system-theoretic point of view, and some idea of "minimality" for proofs of theorems. (We mathematicians think that "mathematical elegance" has something to do with this minimality!)

We may think of a proof as having a combinatorial structure associated with a directed graph. In situations where the semantics of the program is a piece of mathematics, this graph structure must be intrinsically linked to the underlying mathematics!

The computer scientists, when thinking in general terms about the meaning of recursion and iteration, have mainly been oriented towards the number-theoretic and classical logic (Godel, Post, Church, Turing, Kleene, . . .) origins of the concepts. Dana Scott has been a pioneer in suggesting how these classical ideas may be broadened and extended.

However, I also have a broader, long-range agenda in mind. What we in mathematics call LIE THEORY has dominated my own research career in pure and applied mathematics. It deals with the interrelation between differential equations, differential geometry, and certain algebraic structures more-or-less related to groups. Lie theory was begun in the period 1875 - 1925 in the work of Lie and Elie Cartan, and was developed from 1950 to 1965 and by researchers centered around Charles Ehresmann in Strasbourg and Paris, and Donald C. Spencer in Princeton. In recent years, many of their ideas have been applied in such physical and engineering disciplines as elementary particle physics, continuum and particle mechanics, and control theory. However, until now, Lie theory and differential geometry has had no impact or even connection with computer science. My aim is to begin to develop such links.

2. ITERATION AND RECURSION

Iteration and recursion are two mathematical concepts that are fundamental to computer science. Here is one way that we shall think of them.

Let X be a space, and let f: $X \to X$ be a map from X to itself. A <u>fixed point</u> of f is a point $x' \in X$ such that

$$f(x') = x' \quad . \tag{2.1}$$

By <u>recursion</u> we will mean the definition (of a function, computer program, logical proof, . . .) as a fixed point.   Here is one way of developing such a concept.

Given a point $x_0 \in X$, set:

$$
\begin{aligned}
x_1 &= f\!\left(x_o\right) \\
x_2 &= f\!\left(x_1\right) \\
&\;\;\vdots \\
x_n &= f\!\left(x_{n+1}\right) \quad .
\end{aligned}
\tag{2.2}
$$

DEFINITION. The fixed point x' of f is defined <u>via</u> <u>iteration</u> if

$$x' = \lim_{n\to\infty} x_n = \lim_{n\to\infty} f\!\left(\ldots\left(f(x_o)\right)\ldots\right) \quad . \tag{2.3}$$

For the moment, we leave open the precise definition of "limit" on the right-hand side of (2.3).

We do have the following well-known result.

THEOREM 2.1. Suppose X has a topological structure such that the map $f\colon X \to X$ is continuous. Let f^n be the n-fold iterate of f. Let x be a point of X. Set $x_n = f^n(x_0)$. Suppose that:

$$\lim_{n\to\infty} x_n = x \quad . \tag{2.4}$$

Then,

$$f(x') = x \quad . \tag{2.5}$$

PROOF.

$$f(x) = f\left(\lim_{n \to \infty} x_n\right) = \quad , \quad \text{by continuity of } f,$$

$$\lim_{n \to \infty} f(x_n) = \lim_{n \to \infty} x_{n+1} = x \quad .$$

Q.E.D.

One of Scott's basic ideas is to turn this reasoning around, and attempt to define a topological structure for X such that the map f becomes continuous, so that (2.4) is satisfied, and so that the neighborhood structure on X associated with the topology is linked with intuitive and/or logical notions of "computability." However, I will not work here within the lattice-theoretic and categorical framework of Scott's papers, but will try to work in a concrete setting more immediately appropriate for the applications to differential and difference equations.

Let us see how this works for a standard example in computer science programming and language texts, the factorial function on the integers.

3. THE FACTORIAL IN TERMS OF RECURSION AND ITERATION

Let

$$N = \left\{ n: \ n = 1, 2, \ldots \right\} \quad , \tag{3.1}$$

i.e., N is the space of positive integers. Let

$$X' = \text{space of maps } x: \ N \to N \text{ such that } x(2) = 1 \quad . \tag{3.2}$$

Define a map f: $X' \to X'$ as follows:

$$f(x)\,(n) = n\,x(n - 1) \quad \text{if } n > 1 \tag{3.3}$$

$$= 1 \quad \text{if } n = 1 \ . \tag{3.4}$$

THEOREM 3.1. There is exactly one fixed point of the mapping f. It is the factorial function:

$$x(n) = n! = n(n - 1) \ldots 1 \ . \tag{3.5}$$

PROOF. x is a fixed point of f if f:

$$f(x)\,(n) = x(n) \quad \text{for all } n \in R \ . \tag{3.6}$$

(3.6) means that:

$$x(n) = n\,x(n - 1) \ . \tag{3.7}$$

(3.7) is, of course, a difference equation. $n \rightarrow x(n) = n!$, as a map $N \, \text{Æ} \, N$, is a solution.

To verify that (3.5) is, in fact, a solution of the difference equation (3.7), and that it is unique subject to the initial condition $x(2) = 1$, is an easy proof by induction, left to the reader.

$$\textbf{Q.E.D.}$$

In order to construct $n \rightarrow n!$ iteratively   set:

$$X' = \big\{ (n, x)\colon \ n \text{ is a positive integer (possibly infinite)}, \ x \text{ is a map of}$$

$$N \cap [2, n] \text{ into } N, \text{ and } x(2) = 1 \big\} \ . \tag{3.7}$$

X is then a subset of X'. $x \in X$ is identified with the point (∞, x) of X'.

Define a map f: $X' \to X'$ as follows:

> If x is a map $N \ll [2, n] \to N$, then f(n, x) is the map $N \ll [2, n + 2] \to N$ defined as follows:

$$f(x)\,(2) = 1$$

$$(3.8)$$

$$f(x)\,(n) = n\,x(n - 1) \quad \text{for } n > 1 \ .$$

DEFINITION. Let (n, x), (n', x') be two elements of X'. Then

$$(n, x) \le (n', x') \tag{3.9}$$

if and only if the following conditions are satisfied:

$$n \le n' \ , \tag{3.10}$$

$$x(n) = x'(n) \quad \text{for all } n \in [1, n] \ . \tag{3.11}$$

(It is readily verified that (3.10) - (3.11) defines "$\le$" as a <u>partial ordering</u> on X', thus justifying the notation.)

THEOREM 3.2. The map f: $X' \to X'$ defined by (3.2) is monotone with respect to the partial ordering (3.9) - (3.11); i.e., we have:

$$f(n, x) \le f(n', x') \ . \tag{3.12}$$

PROOF. (3.13) follows routinely from (3.8).

$$\text{Q.E.D.}$$

THEOREM 3.3. Let x_0 be the "base" element of X'; i.e., $x_0(1) = 1$. Set:

$$x_n = f^n(x_o), \quad n > 1 \quad . \tag{3.13}$$

Let x be the following element of X':

$$x(n) = n! \quad \text{for all } n \in N \quad . \tag{3.14}$$

Then,

$$x = \lim_{n \to \infty} x_n \quad . \tag{3.15}$$

Here, one way of interpreting what "limit" on the right-hand side of (3.15) means is that the following condition is satisfied:

$$\text{For each } n \in N, \, x(n) = x_m(n) \text{ for m sufficiently large} \quad . \tag{3.16}$$

4. DEFINING ITERATIVE SOLUTIONS OF RECURSION EQUATIONS

Suppose that X is a set and that $x' \in X$ is a point of X for which we want to devise a computational "strategy" or algorithm. As an example to keep in mind, think of the definition of the factorial function :   N → N by recursion/iteration. In that case, X is the space of partial functions : N → N.

Let us suppose that f: X → X is a map such that f(x') = x'. In Section 2, we have discussed conditions of a topological nature which assured us that x' could be calculated via iteration as follows:

$$x' = \lim_{n \to \infty} f^n(x_1) \, , \quad \text{for some } x_1 \in X \quad . \tag{4.1}$$

However, as we have seen in the example of Section 2, in the mathematical structure called a 'partial ordering', it offers possibilities for defining elements via iteration. Let us recall the following:

DEFINITION. A **partial ordering** on the set X is given by means of a partially defined binary relation $(x_1, x_2) \rightarrow x_1 \leq x_2$ defined between pairs of points of X satisfying the following rules:

$$x_1 \leq x \quad \text{for all } x \in X \ , \tag{4.2}$$

$$x_1 \leq x_2 \quad \text{and} \quad x_2 \leq x_3 \quad \text{implies} \quad x_1 \leq x_3 \ , \tag{4.3}$$

$$x_1 \leq x_2 \quad \text{and} \quad x_2 \leq x_1 \quad \text{implies} \quad x_1 = x_2 \ . \tag{4.4}$$

Having defined such an algebraic relation at the set-theoretic level, we can also define the analogue of "homomorphism."

DEFINITION. Let X, X' be sets, with partial orderings given on each. A map f: $X \rightarrow X'$ is said to be **monotone** if the following condition is satisfied:

$$f(x_1) \leq f(x_2) \quad \text{for all pairs } (x_1, x_2) \in X \times X \text{ such that}$$

$$x_1 \leq x_2 \ . \tag{4.5}$$

THEOREM 4.1. Suppose that X is a partially ordered set, and that f: $X \rightarrow X$ is a map that is monotone, in the sense that (4.5) is satisfied. Let x_0 be a point of X such that

$$x_0 \leq f(x_0) \ . \tag{4.6}$$

Let $\{x_n: n \geq 1\}$ be the sequence of points of X defined inductively as follows:

$$x_1 = f(0)$$
$$\vdots \qquad\qquad (4.7)$$
$$x_{n+1} = f(x_n) \ .$$

(In other words, x_0, x_1, x_2, ... , is the orbit at x_0 of the semigroup generated by f acting on X.) Let x' be a point of X satisfying the following equation:

$$x' = \text{least upper bound of } \{ x_n: n = 1, 2 ... \} \ .$$

Then,

$$f(x') \geq x' \ . \qquad\qquad (4.8)$$

PROOF. Apply f to both sides of (4.6) and use its monotonicity.

$$f(x_0) = x_1 \leq f(x_1) = x_2 \ . \qquad\qquad (4.9)$$

Applying f iteratively to (4.9), we have

$$x_n \leq x_{n+1} \ .$$

To say that "x' is a least upper bound of the set $\{x_n\}$" is to say that the following two conditions are satisfied:

$$x_n \leq x' \ \text{ for all n } , \qquad\qquad (4.10)$$

$$x' \leq x'' \quad \text{for all } x'' \in X$$

$$\text{such that } x_n \leq x'' \quad \text{for all } n \ . \tag{4.11}$$

In particular, notice that x' is the unique element satisfying these conditions.

Apply f to both sides of (4.11) and apply monotonicity again:

$$f(x_n) = x_{n+1} \leq f(x') \ . \tag{4.12}$$

(4.12) implies that f(x') is also an upper bound; hence, f(x') $\geq$ x'.

Q.E.D.

Let us now find reasonable sufficient conditions that the inequality (4.8) can be reversed. Now, we have the following chain of inequalities:

$$f(x_0) \leq f(x_1) \leq \ldots \leq f(x_n) \leq \ldots \leq x' \ . \tag{4.12}$$

Now, if X' were the real numbers, with the ordering just that given naturally, then from (4.12) we could deduce from continuity of f that

$$f(x') \leq x' \ ; \tag{4.13}$$

i.e., the reverse of (4.8). (4.8) and (4.13) would imply that

$$f(x') = x' \ ;$$

i.e., x' is a fixed point of f.

In Scott's work, the reader will find an extensive discussion of when (4.13) is satisfied. This has led to the theory of <u>continuous lattices</u>. Since I do not want to detour to cover this topic in its own generality, I will deal only with the most important case for computer science and logic, the case where X is a space of partial functions.

5. SOLUTION OF FIXED POINT EQUATIONS FOR MAPPINGS ON SPACES OF PARTIAL FUNCTIONS

Let Y and Z be sets and let:

$$X = \text{space of partial maps } Y \to Z \ . \tag{5.1}$$

Scott uses the following notation.

$$X = (Y \to Z) \ . \tag{5.2}$$

I find it to be an excellent notation, and will use it myself.

Adopt the following partial ordering on X:

$x1 \le x2$ if and only if the following conditions are satisfied:

$$\text{Domain}\left(x_1\right) \subset \text{Domain}\left(x_2\right) \ , \tag{5.3}$$

$$x_2(y) = x_1(y) \ , \tag{5.4}$$

for all $y \in$ Domain $x1$.

Let f: $X \to X$ be a map. Let x_0 be an element of X such that

$$x_o \le f\left(x_o\right) \le f\left(f(x_o)\right) \le \dots \ . \tag{5.5}$$

Suppose that the monotone sequence in X on the left-hand side of (5.5) has a least upper bound x'. By (4.8), we have

$$f(x') \geq x' \quad . \tag{5.6}$$

We now have the following result.

THEOREM 5.1. Suppose that the following condition is satisfied:

$$\text{Domain } (x') = Y \quad . \tag{5.7}$$

Then,

$$f(x') = x' \quad ;$$

i.e., x' is a fixed point of f.

PROOF. If (5.7) is satisfied, the reverse inequality to (5.6) follows automatically from the definition (5.3) - (5.4) on the space of partial maps.

Q.E.D.

REMARK. This result is very suggestive from the computational point of view. Think of the sequence of partial functions

$$x_o \leq f(x_o) \leq \ldots f^n(x_o) \leq \ldots$$

as the result of applying the algorithm or program f iteratively to the initial data x_o. Then, condition (   ) may be interpreted as meaning that the algorithm f terminates "successfully" in the computation of the <u>total</u> function x', which will then provide a "solution" to the recursion equation

$$f(x') = x' \quad .$$

It seems to me that a major goal of control theorists (and other mathematical engineering disciplines that are closely linked to the traditional differential-equations way of looking at the world) in the coming years must be to develop such a "recursive" -- and computer implementable -- way of looking at their discipline. A first result of such a re-orientation might be a more determined effort to reconcile numerical analysis which is traditionally oriented towards the differential equations of continuum mechanics -- and the more analytically oriented problems of control.

With this motivation as background, let us turn to a topic from differential equation theory that I believe can play the same foundational role that the simple number - theoretic difference equations (such as the defining factorial and Fibonacci numbers) have historically played in computer science thinking. However, we encounter a fundamental computer science theoretic difference between differential and difference equations: The general algorithms for the solution of differential equations involve infinite processes (such as the Picard iteration algorithm) which are not exactly implementable on computers. Of course, approximations are feasible; and indeed the contemporary theory of the numerical analysis of ordinary differential equations is dominated by the need to "modernize" the classical, hand-computation-oriented algorithms to adapt them to contemporary computers. Here, certain questions of efficiency and robustness of approximation become most urgent.

As an illustration of how such approximations of fixed point algorithms might be structured, let us turn to study the standard existence algorithm for ordinary differential equations.

6. APPROXIMATION OF THE PICARD ITERATION ALGORITHM FOR ORDINARY DIFFERENTIAL EQUATIONS

The Picard iteration procedure is a classical example of an iteration procedure that provides an existence proof. Let us recall its formulation.

Consider a system of n, first order ordinary differential equations; say, of the following form:

$$\frac{dx_1}{dt} = f_1(x_1, \ldots, x_n)$$
$$\vdots$$
$$\frac{dx_n}{dt} = f_n(x_1, \ldots, x_n) \ . \tag{6.1}$$

Let us write the system (6.1) in the following vectorial form:

$$\frac{dx}{dt} = f(x) \ ,$$

$$x = (x_1, \ldots, x_n) \ , \tag{6.2}$$

$$f = (f_1, \ldots, f_n) \ .$$

x then denotes an element of R^n; f a map

$$f: R^n \to R^n \ . \tag{6.3}$$

A solution of (6.3) is a map

$$x: t \to x(t) \tag{6.4}$$

of $R \to R^n$ satisfying (6.2). Integrating both sides of (6.2) over the interval

$$0 \leq s \leq t \quad \text{of real numbers} \tag{6.4}$$

appears as the solution of the following integral equation:

$$x(t) = x(0) + \int_0^t f\big(x(s)\big) \, ds \ . \tag{6.5}$$

We can interpret (6.5) as a fixed point condition. Given the element:

$$x_0 \in R^n \; ,$$

define a map

$$P_f \colon \left(R \to R^n\right) \to \left(R \to R^n\right)$$

as follows:

$$P_f(\mathbf{x})\,(t) = x_0 + \int_0^t f\big(x(s)\big)\,ds \quad . \tag{6.6}$$

An element

$$\mathbf{x} \in \left(R \to R^n\right)$$

such that

$$P_f(\mathbf{x}) = \mathbf{x} \tag{6.7}$$

then corresponds to a solution of (6.5) (or (6.2)) satisfying the following initial condition:

$$\mathbf{x}(0) = x_0 \quad . \tag{6.8}$$

It is well-known that under certain conditions a solution to (6.7) may be found by starting with

$$\mathbf{x}_0 = \text{constant function } t \to \mathbf{x}_0 \ ,$$

and iterating it:

$$\mathbf{x}_1 = P_f(\mathbf{x}_0)$$

$$\mathbf{x}_2 = P_f(\mathbf{x}_1)$$

$$\vdots \qquad\qquad (6.9)$$

$$\mathbf{x} = \lim_{j \to \infty} \mathbf{x}_j \ ,$$

Numerical analysis is concerned with computational algorithms and strategies which approximate this procedure, with the emphasis on those that are effective in terms of FORTRAN programs for today's computers. Here, I want to think about more "geometric" procedures for studying approximate algorithms.

The smallest way to approximate the integral operator Pf is to use the Riemann Sum approximation for the integral on the right-hand side of (6.6). First, change the range of integration from $0 \le s \le t$ to $0 \le s \le 1$:

$$P_f(x)\,(t) = x_0 + t \int_0^1 f\big(x(ts)\big)\, ds \ . \qquad (6.7)$$

As an illustration of what it means to "approximate" the operator Pf, let us use the following approximation to the integral:

$$\int_0^1 g(s)\, ds \sim \sum_{j=0}^{n-1} g\left(\frac{j}{n}\right)\frac{1}{n} \ , \quad n = 1, 2, 3, \ldots \ . \qquad (6.8)$$

This leads to:

$$P_f(x) \sim x_0 + \frac{t}{n} \sum_{j=0}^{n-1} f\left(x\left(\frac{tj}{n}\right)\right) \ . \tag{6.9}$$

This suggests defining the following approximate operator:

$$P_{t,n}(x)\,(t) = x_0 + \frac{t}{n} \sum_{j=0}^{n-1} f\left(x\left(\frac{tj}{n}\right)\right) \ . \tag{6.10}$$

THEOREM 6.1. For each integer n, formula (6.10) defines P_f as a map:

$$\left(R \to R^n\right) \to \left(R \to R^n\right) \ .$$

An element $t \to x(t)$ of $(R \to R^n)$ is a fixed point of P_f^n if and only if it satisfies the following difference equation:

$$x(t) = x_0 + \frac{t}{n} \left[f\left(x\left(\frac{(n-1)\,t}{n}\right)\right) + \ldots + f\left(x(0)\right) \right] \ . \tag{6.11}$$

This difference equation is a example of those classified in the standard treatisesas linear multistep methods for numerical solution of ordinary differential equations. Sufficient conditions that the solution of (6.11) converges as $n \to \infty$ pointwise in t are available in the numerical analysis literature . Of course, this is the simplest of such methods, the so-called Euler method.

Let us now turn to a more Lie - theoretic methodology.

7. THE FEYNMAN APPROACH TO ODE APPROXIMATION AND REPRESENTATION

Let us return for a moment to our goal of linking together the computer and differential equations view of phenomena. Since, as a mathematical function, a "differential equation" (or rather, the space of its solutions) is not a finitely computable object -- that is, it cannot be described by means of a finite-length computer program acting on the language of symbols and numbers available on standard computers -- it is necessary to think of approximate means. In this section, I will describe one method of thinking about approximations to differential equations, based on Feynman's famous work on path integral solutions to the differential equations of quantum mechanics. What I find attractive in Feynman's point of view is that it is based at least implicitly on the Lie - groups interpretation of differential equations.

Let us now think of the system (6.1) of ordinary differential equations from the Lie theoretic point of view . Set:

$$v = f_1(x_1, \ldots, x_n) \frac{\partial}{\partial x_1} + \ldots + f_n(x_1, \ldots, x_n) \frac{\partial}{\partial x_n} \ . \qquad (7.1)$$

For algebraic convenience, we shall suppose that the real-valued functions $f_1, \ldots, f_n$ on R^n that appear on the right-hand side of (7.1) are C^∞ ("infinitely differentiable") functions. The first-order linear partial differential operator v may then be interpreted as a vector field on the manifold R^n; i.e., as a cross-section of the tangent vector bundle of R^n.

DEFINITION. $F(R^n)$ denotes the subset of maps $(R^n \to R)$ that are infinitely differentiable in the calculus sense. We shall consider $F(R^n)$ as a commutative, associative algebra, with the binary algebra operations

$$(f_1, f_2) \to f_1 + f_2$$

$$\left(f_1\, f_2\right) \to f_1\, f_2 \ ,$$

those given by pointwise algebraic operations on the values of functions; i.e.,

$$\left(f_1 + f_2\right)(x) = f_1(x) + f_2(x)$$

$$\left(f_1\, f_2\right)(x) = f_1(x)\, f_2(x)$$

$$\text{for } f_1, f_2 \in F\left(R^n\right)$$

$$x \in R^n \ .$$

(7.2)

We can think of a first-order linear differential operator of type (6.2) as a derivation of the algebra $F(R^n)$.

$$f \to f_1 \frac{\partial f}{\partial x_1} + \ldots + f_n \frac{\partial f}{\partial x_n}$$

Denote the space of such derivations as $V(R^n)$.

A curve $t \to x(t)$ in R^n is an orbit curve of the vector field V if it satisfies the following condition:

$$\frac{d}{dt}\, f\bigl(x(t)\bigr) = v(f)\bigl(x(t)\bigr) \ ,$$

$$\text{for all } f \in F\left(R^n\right) \ ,$$

(7.3)

$$t \in R \ .$$

THEOREM 7.1. Suppose that the curve $t \to x(t)$ satisfies condition (7.3). Then, for each $a \in R$, the curves

$$t \to x(t + a) \qquad (7.4)$$

also satisfy equations (7.3). Conversely, conditions (7.5) are sufficient for the curve $t \to x(t)$ to satisfy (7.3).

PROOF. For $f \in F(R^n)$,

$$\frac{d}{dt} f\big(x(t+a)\big) = \frac{df}{d(t+a)} \frac{d(t+a)}{dt} = v(f)\big(x(t+a)\big) \bullet 1 \ .$$

Here (7.4) also satisfies (7.3).

Q.E.D.

THEOREM 7.2. With the vector field $V \in V(R^n)$ given by formula (7.2), the curve $t \to x(t)$ satisfies condition (7.3) if and only if it satisfies the following system of first-order ordinary differential equations:

$$\frac{dx_1}{dt} = f_1\big(x_1(t), \dots, x_n(t)\big)$$

$$\vdots \qquad\qquad (7.5)$$

$$\frac{dx_n}{dt} = f_n\big(x_1(t), \dots, x_n(t)\big) \ .$$

PROOF. The coordinate functions

$$\big(x_1, \dots, x_n\big) \to x_2, \dots, x_n$$

are elements of $F(R^n)$. Substituting them in turn for f in (7.3) gives relation (7.5). This proves Theorem 7.2 one way.

For the converse, suppose that $t \to x(t)$ is a curve in R^n that satisfies (7.5). Let $f \in F(R^n)$. Then, f is a function of the coordinate functions

$$x_1, \ldots, x_n. \text{ Hence, for } f \in F(R^n),$$

$$\frac{d}{dt} f(x(t)) = \sum_{j=1}^{n} \frac{\partial f}{\partial x_i} \frac{dx_i}{dt} = \sum_{j=1}^{n} f_i(x(t)) \frac{\partial f}{\partial x_i} = v(f)(x(t)) \ .$$

Q.E.D.

Let us specialize (7.5) to be a linear time-independent system that we write in the following vectorial form:

$$\frac{dx}{dt} = Ax \ , \quad x \in R^n \ . \tag{7.5}$$

A is a linear map: $R^n \to R^n$ realized by an n x n real matrix. Then, we can write the solution curves $t \to x(t)$ of the system of differential equations (7.5) as orbits of the one-parameter group

$$t \to \exp(tA) = \sum_{j=0}^{\infty} \frac{t^j A^j}{j!} \tag{7.6}$$

of matrices:

$$x(t) = \exp(tA)(x(0)) \ . \tag{7.7}$$

For each $t \in R$, $\exp(tA)$ determines a diffeomorphism of $R^n \to R^n$, and

$$\exp(t_1 + t_2)(A)(x) = \exp(t_1 A) \exp(t_2 A)(x) \ ,$$

$$t_1, t_2 \in R \ , \tag{7.8}$$

$$x \in R^n \ .$$

Relation (7.8) means that $t \rightarrow \exp(t\,v)$ is a one-parameter group of diffeomorphisms of R^n.

It will be useful to write the relations between the orbit curves of the (in general, nonlinear) vector field v determined by equations (7.3) or (7.5). In an analogous form,

$$x(t) = \exp(t\,v)\big(x(0)\big) \tag{7.9}$$

where $t \rightarrow \exp(t\,v)$ is a one-parameter group of diffeomorphisms of R^n generated by the vector field v. (Strictly speaking, $t \rightarrow \exp(t\,v)$ is only a one-parameter pseudogroup of local diffeomorphisms. However, we shall not explicitly take into account here this possible complicating feature of the discussion.)

The analogue of the power series definition (7.6) of the "general solution" of a linear equation is the following Lie series expansion:

$$\exp(t\,v)^k (f) = \sum_{j=0}^{\infty} \frac{t^j}{j!} v^j (f), \quad \text{for } f \in v(R^n) \ . \tag{7.9}$$

This group - theoretic way of describing the relation between the totality of solutions of an ordinary differential equation and its vector field infinitesimal generator is the basis of Lie theory. We shall use it also to describe how to find approximations to the solutions of (7.5). Feynman's famous (and Nobel-prize-winning work) is based on extensions of these ideas to treat the one-parameter groups which act in quantum mechanics.

Suppose that

$$t \rightarrow x(t) = \exp(t\,v)\big(x(0)\big)$$

is a curve in R^n that is an orbit curve of the vector field v, as explained above. First, using the group property (7.8), we have the following result.

THEOREM 7.3. Let $a_1, \ldots, a_n$ be a sequence of real numbers such that

$$a_1 + \ldots + a_n = 1 \ . \tag{7.10}$$

Then, let $t \to x(t)$ be an orbit curve of the vector field V.

Set:

$$g(t, x) = \exp{(t\,v)}\,(x) \ , \quad \text{for } t \in R \ , \quad x \in R^n \ . \tag{7.11}$$

Then, we have the following relations:

$$x(t) = \exp\left(a_n\,t\,v\right) \ldots \exp\left(a_j\,t\,v\right)\left(x(0)\right) \ , \tag{7.12}$$

$$= \exp\left(a_n\,t\,v\right) \ldots \exp\left(a_j\,t\,v\right) g\left(a_j\,t\,x(0)\right) \ , \tag{7.13}$$

$$= \exp\left(a_n\,t\,v\right) \ldots g\left(a_2\,t\right) g\left(a_1\,t\right)\left(x(0)\right) \ . \tag{7.14}$$

$$\vdots$$

These formulas offer what in computer science is called an iterative description of the solution. The corresponding recursive description is that the solution curves $t \to x(t)$ are fixed points of the one-parameter group (7.11) in the sense that the following relation is satisfied:

$$g\left(t, x(s)\right) = x\left(t + s\right) \ , \quad \text{for } t, s \in R \ . \tag{7.15}$$

PROOF. Relations (7.12) - (7.15) follow immediately from the group relation (7.8) and the postulated condition (7.10).

Q.E.D.

We can now describe what we shall call the **Feynman** strategy for approximating solutions of (7.5).

HEURISTIC PRINCIPLE

Suppose that

$$(t, x) \to h(t, x) \tag{7.16}$$

is a mapping of $R \times R^n \to R^n$ satisfying the following condition for fixed $x \in R^n$:

$$h(t, x) \text{ is approximately } g(t)(x) = \exp(t\,v)(x) \ . \tag{7.17}$$

Then, for fixed $t \in R$, and for $a_1, {}^\circ\,, a_n$ chosen to be "small" relative to one, the solution $t \to x(t)$ of the differential equations (7.5) satisfying the following initial condition:

$$x(a) = x_o \tag{7.17}$$

is given by the following approximate iterative formula.

$$x(t) \sim h\!\left(a_m\, t\right) h\!\left(a_{m-1}\, t\right) \ldots h\!\left(a_1\, t\right) x(0) \ . \tag{7.17}$$

PROOF. Substitute the relation

$$h(t, x) \sim g(t)(x) \quad \text{for } 0 < t \ll 1$$

into (7.14).

REMARKS ON AI AND CONTROL

There have been several Workshop at NASA-Ames searching for the common ground beteen control and AI. The first, in Robotics, taught there was a new world of application, combining concepts familiar to us from the background in the mathematics of Control and ideas central to Computer Science. We learned that Computation is the central issue in any application of computer science: In Control Theory, with its traditional methodology based on differential equations, Symbolic Computation - the subject of the second Workshop - is central. The thirdWorkshop featured topics from AI and that part of Control in closest symbiosis with CS, Discrete Event Sytstems.

As a result of this learning-through-doing experience, I am very encouraged that there is much to be gained - both from the point of view of Technology and Science - by researchers in both CS and Control from pooling talents and research ideas. AI has, from its birth, of course been closely tied to the advance of computer hardware and software: Indeed, much of the progress in these areas for the past thirty years has been made by those rising to the challenge laid down by the pioneers of AI, to understand Intelligence as Computation, and to simulate as much as feasible the mysterious workings of the Mind.

For the past several years, I have studied Computer Science and its sub-discipline Artificial Intelligence with the aim of finding applications to Engineering. In order to achieve maximum benefit from existing computer science and technology, I believe that much more work at a fundamental level is necessary: What the AI people call the 'Formal' side of the subject. Above all, we must think much more algebraically about the fundamentals of Computer Science. There has already been extensive application of algebraic methods to the theory of semantics of computer programs: I have in mind the extension of some of these methods to the analysis and design of the more applied side of Computer Science, and to the process of Programming itself. There has recently been talk about revitalization of scientific computation, particularly from the point of view of extending the methodology from the traditional numerical, FORTRAN -based methods to newer, mixed symbolic-numerical possibilities inherent in such languages as LISP and PROLOG.

Control Theory offers excellent opportunities here, since it is a part of applied mathematics that is closely linked to important parts of core mathematics, especially algebra and differential geometry. Control also plays an imprtant role in Robotics. There are control-theoretic components of

Robotics that   offer magnificent opportunities for the extensions and applications of existing Control and CS theory and technology.

Most of the mathematics-related effort by computer science-oriented and trained researchers has approached the problem of achieving goals via the intellectual tools provided by logic. While there are certainly sound reasons for this emphasis - and historical ones as well, given the close links between logic, language and computer program theory - I believe that sufficient emphasis has not been placed on the possibilities of algebrazation of the process. A pioneering work by Wu-Wen Tsun - codified in a recent book by Chou - on theorem proving in geometry offers glimpses of what can be done by thinking more algebraically about the mathematical methodology of computer science.

One barrier to more effective interaction and communication between the CS/AI communities and the broad spectrum of the active pure-and applied mathematics researchers has been the need for the former to think in a relatively narrow, constrained way. This barrier can be broken down by each side moving towards the other: CS/AI people - especially the students at the major research institutions - should become familiar with a broader spectrum of mathematics, and the mathematical world should study more deeply and thoroughly the point of view of contemporary CS: Data Structures, Algorithms, Recursive and Constructive Methods,.... . Here I am struck by historical analogies - in fact from my own experience - with what has happened in Elementary Particle Physics. When I first encountered this discipline twenty-five years ago, it had developed - in a severely'pragmatic' way - its own methodology and language, partly in Oedipal rejection of the path laid out by Einstein. However, in the early 1960's the theory of semi-simple Lie groups created by Lie, Cartan and Weyl took a dominant role in the thinking of the leading physicsts, and the subject has - with great success - become more and more interwined with deep mathematical questions and structure, but without losing its primary orientation towards practice and experiment. Reading how mathematics is treated in the books that play the leading role in educating the current generation of researchers in CS and AI reminds me in a very striking way of the analogous situation in Physics in the early 1960's.

As part of my job of thinking about how the NASA effort may be helped by CS/AI, I have spent considerable time over the past four years studying and surveying for mathematical content the bewildering variety of CS/AI theories and languages. It seems to me that the SCHEME dialect of LISP - as expounded in the textbook/treatise by Abelson and Sussman - offers the best glimpse of future possibilities of the development of tools that will bring much closer together the 'working' research mathematician and the computer scientist, in much the same way that elemetary particle physics and differential geometry have interacted. Of course, I do not expect identity - or even much explicit convergence or coincidence of research goals - but rather a more subtle and indirect mutual interaction. The first step towards this might be the development of a collection of research-oriented books and survey papers exploring the common ground. (I have a bias here, since I spent a good portion of my own research life in the 1960's writing such material for the potential benefit of the Elementary Particle and Differential Geometry/Lie Groups communitities. I should warn anyone thinking of doing the job I am suggesting here for the much-broader CS/AI culture that it is, in my experience, a thankless task!)

The point that I see to begin the development of this common ground is the seminal work of John McCarthy and Dana Scott (both students of mathematics at Princeton in the 1950's   : a coincidence?) who recognized - and made fundamental contributions to - the importance of the development of **recursive** ways of thinking about mathematics and CS. It seems that further development of the directions pioneered by Scott from a combined CS and mathematical point of view has been inhibited by the lack of mutual interaction and comprehension between the two cultures. For example, he has discovered remarkable and deep contact between recursion theory and such mathematical disciplines as Lattice Theory, Topology, and Category Theory, but one can find very little of this embedded in the CS research community. Perhaps the CS culture has been so dominated by the combinatorial nature of the main problems of the past thirty years that they have not been able to teach their students a lesson we mathematicians learned to our great benefit, that one cannot go into a new problem area and say that, for example, one is only going to think about a certain restricted piece of mathematics!

One of the features of important 'applied' mathematics is that it forces a re-thinking - and possibly even a re-alignment - of the 'pure' aspects as well. This feature may be regarded as the equivalent in mathematics of the interplay between 'theory' and 'experiment' that is so important and strikingly and characteristically frutiful in Physics.

The emergence of CS in the past twenty year has closely involved two areas of mathematics: Combinatorics and mathematical logic. Thirty years ago, both were in back-waters of the evolutionary mainstream of mathematics: That they have emerged so strikingly is partially due to their utility in CS. (Of course, there have also been developments internal to the disciplines that has contributed to this rejuvenation.) However, there still seems considerable incomprehension between CS and the more mainstream parts of mathematics. I hope that some funding agency that is interested in making a serious effort will take this up as some sort of 'mission', and foster the development of a full-spectrum of interaction.

My study of CS in the past four years has had the beneficial side-effect of forcing me to study the logic and foundations literature, which I admit I have never found particularly congenial. (I am probably typical in this attitude of most mathematicians of my generation.) I am an adherent of the 'mathematical structures' philosophy, most closely associated with the Bourbaki schoool. I have spent most of my life as a 'working' mathematician pursuing the applied ramifiactions of two such structures: Lie groups and algebras, and differentiable manifold theory. What I found initially disconcerting about the computer science approach was the need to begin at the beginning, with *language*, and the process of deduction from that primitive a beginning made very tediously explicit. Of course, this is the intellectual method of the logicians, and I have slowly overcome my own mathematical roots and begun to understand and be comfortable with this way of thinking.

As a differential geometer, *differential equations* have been fundamental to my way of thinking about the world. The equivalents to the the physicist and engineer are *physical laws*, which of course are often codified in terme of differential equations. Part of the difficulty in reasoning about the physical world that one encounters in the Robotics and AI literature is that the miraculous powers conferred by this way of thinking and mathematizing are substantially lost.

I believe that the theory of scientific computation is important also for reasons internal to CS. It seems to me that this young and vigourous discipline has been excessively been polarized between the mundane and the inspirational. On the one hand, most of the work in this initial thirty-five years or so has been oriented towards the bread-and -butter data processing problems that pay the bills. (Of course, I am not denigrating this aspect: The development of every scientific discipline requires a stock of such experience, not the least mathematics!, and it is healthy.) On the other hand, the inspiration has often come from the mish-mash grandeloquently called Artificial Intelligence. Again, much of benefit to everyone has come from this, especially the goal of developing more powerful and sophisticated languages such as LISP and PROLOG, the importance of mechanizing Compuational Logic and Knowledge Representation, extending Mathematical Logic to cover more realistic portions of the world, and so on. Strangely, the field of Scientific Computation, and within that the problem of mechanizing mathematical reasoning (not only Theorem Proving, as it has developed within Logic!) has received   short-shrift. (The only exception - and of course it is a major one - perhaps the one that proves the rule! - is the deveolpment of MACSYMA.) Since I am an outsider, I can only guess at the reasons for this. Certainly it is an illustration of the Law of Scientific Funding that it is unhealthy for most of the money for a discipline to come from only one source!

Within the broad field of Scientific Computation, my own special field of interest is in the ways of reasoning about differential and difference equations. Historically, there has been somewhat of a dichotomy between ways of thinking about the world exemplified by Mathematical Logic and Differential Equations/Scientific Laws. The AI culture has increasingly approached this gap and tried to bridge it with such attempts as Qualitative Physics and Temporal Logic. The latter subject - technically a branch of Modal Logic - originated as a method of thinking about the sort of reasoning that goes on in checking the correctness of a computer program: The "time" in this situation is the integer that labels the sequence of instructions that are fed into a program for a von Neumann-type of computer, and there are attempts in the contemporary AI and CS literature to apply these techniques to broader circumstances. The importance for engineering purposes of even modest success in this direction is obvious to everyone, but I believe that there are major complexity problems to be conquered before the applications materialize. Perhaps it will turn out that the reason we were given Differential Equations to play with is that they are a great tool for achieving certain goals while keeping compuational complexity under control!

I have studied the Programming Theory and AI literature, particularly the work on Recursive Functions, Default and Modal Logic and attempts ('Geometric Logic') to apply to Logic reasoning originating in Algebraic and Differential Geometry . Let me review a few common general themes: Conventionally, a Computer Programming Theory consists of three components: First, a **language**, i.e. a set of symbols and rules for combining them. Second, a **syntax**, which picks out admissable collections of these symbols. Third, a **semantics** which assigns mathematical objects, such as sets and mappings, to syntactically admissable collections of symbols . Thus 'semantics' acts much like a **functor** from categories that are 'linguistically' defined to the set-theoretic categories needed to describe the physical and engineering world. This over-arching view of the nature of computer programming theory owes much to the world-view of the **logician**. There is a gulf between such a general picture and what must be done in practice to write a program for some computational purpose: It is of course basically the difference between pure and applied mathematics!

Following an analogy suggested to me by Gerald Sussman, think of a Progam defined as above as the analogue of a differential equation. Then think of the computation produced as a result of the program as the analogue of the solution or trajectory of the differential equation. The system of trajectories must be prescribed by the programmer to perform given tasks. Think of mechanics, which has rules (traditionally called 'rational mechanics') for translating the physical and kinematic data into dynamical equations. Unfortunately, there seems to be nothing like a Lagrangian for computer programming theory which will generate the equations for trajectories in some automatic way! Instead, one might look to feedback control theory (and recursive function theory, which in a sense is a general version) for ideas about generating the desired trajectories. Thus we take the view of the 'engineer' rather than the 'physicist' in thinking about the computer environment: The trajectories are to be defined by some process analogous to feedback, subject to such engineering criteria as 'robustness' and 'stability' rather than by the physicist's method of looking for Laws of Nature.

Another way of thinking about this is to say that we want to develop much further - and for 'applied' purposes - the analogies that (at the moment, weakly exist) between the world of Differential Equations that has been so fantastically successful in science and engineering and the world of Recursive Functions that is the historical origin of Computer Science. Ways of minimizing combinatoric complexity dominate the practical computer scientist's world, whereas the traditional methods of Differential Equations have given a way for the engineer to minimize such difficulties. As yet this view emphasizing the analogy between a 'computer program' and a 'differential equation' is in a very primitive state. However, I believe that it is important potentially for what we are trying to achieve in building bridges between the world-view of the engineer/scientist/mathematician and the computer scientist. In a way, it is just a more concrete version of the vision that Scott laid out for us of Category Theory as the unifying theme of Logic and its applications.

At the moment, the following methodology seems necessary when approaching a problem - say, a dynamical system - from the point of view of the Computer Scientist. First comes the Language and Syntax. Then construction of some sort of Data Base methodology. What we particularly need is efficient techniques for describing trajectories of the dynamical system, and ways of choosing the trajectories as efficient as those provided in classical feedback control theory, then ways of changing this data base as new information comes in. Default Logic is, perhaps, one way to think about this. Given the 'complexity' realities, it is probably too idealized a way to deal with the engineering situation.

As a mathematician/geometer who was around at the Beginning of much of contemporary geometry in Princeton and Paris in the 1950's, let me suggest the way I envisage some of the more avant-garde aspects of Geometry - such as Sheaves and Categories - becoming involved in the enterprise of integrating CS and Engineering. (Of course, this Camel's Nose is already under the Tent via Geometric Logic and Algebraic Semantics. However, I am looking for a way to understand its relevance at a more concrete, applied level.). The origin of sheaves is in Leray's work providing an algebraic machinery for extending 'local' to 'global' truth, in the context of cohomological theories. Now, any serious CS structure is going to be a network of algebraic 'things' (Boolean Algebras, Categories, Lattices, Languages, Whatever) hooked up together. A prototype is a Computer Program. Floyd-Hoare Logic - the prototype of much of the Progam Verification literature - involves precisely the Leray idea! A more mathematical model of this might be a system of partial difference equations (in fact, the Leray theory was meant to work with systems of linear partial differential equations: those in CS will be non-linear), with the unknowns having values at the nodes of the network. Thus the solutions are some sort of generalized recursive functions, and statements about the sheaf of solutions can be made using the Leray/Sheaf ideas. In fact, Eilenberg and Elgot (in their 1970 book "Recursiveness") have already given the 'existence proof' that Categories can usefully unify the Recursive Function Theory of Godel, Turing, Kleene, ..., and might be the basis for an important generalization, that one might call Distributed Recursive Function Theory. However, rather that see the proliferation of another Generalized Recursion and/or Algorithm Theory, I would like to see more concrete material demonstrating the usefulness of the general theory - and Recursive Function Theory plays the same role in CS that ODE theory plays in differential geometry - for engineering, mathem,atical and physical practice. For example, I am impressed how in Abelson and Sussman - by far the best thing ever done in the pedagogy of CS! - the logic background plays an unmentioned but important unifying role.

Part of the motivation for an excursion into Logic is the hope that the work on Modal Logic (and 'generalized logics', in general) might prove useful. Now Modal Logic was basically invented for philosophical purposes, and its extension into Computer Science is not automatic. However, it is tempting to pursue the ramifications of Modal Logic. From my point of view, the modal operators are something like correspodences in geometry, but defined for Boolean algebras. (This is basically the point of view in the 1951 Jonsson-Tarski paper). The operator "Eventually something will happen" has the smell of 'asymptotic stability' in 'numerical' control theory. Presumably, a mathematical logician would regard them as an extension of the 'first order language' to describe 'second order things', and wonder about such matters as completeness. Certainly, the computer scientists have written much about this in the context of temporal/dynamic logic.

Now, Peter Caines has been approaching related matters (a 'systems theoretic approach to default logic') via a 'time-varying logic' or a 'time-varying valuation', together with some sort of feedback structure. Matthew Ginsberg's latest paper deals with a many-valued logic approach to default logic that is in fact very close to some sort of 'system theory'. Pratt conceives of Dynamic logic as a sequence of Boolean algebras, together with certain additional structure: This has the smell of a fiber bundle/sheaf whose fibers are Boolean (or maybe Heyting) algebras.

On and on. Clearly there is something going on which benefit from some serious (but hopefully not excessively general and abstract) algebraic treatment! Of course, this puts the goals of ‚naive' AI off into the indefinite future, but this is probably where it belongs! What we should look for is a piece of that pie which can be tied to more concrete problems of science and engineering.

Mathematical logic enters at many points in computer science. (Indeed the cliche that it is "the calculus of computer science" is true!). For computer programming theory, it enters in (at least) three ways. First, it is the foundation for any reasonable theory of semantics of programs. Second, it is essential to any serious theory of Knowledge Representation, and is used there to organize reasoning about the objects described by the program. (However, it is not clear im my mind where 'Knowledge Representation Theory' ends and 'Data Base Theory' begins, so perhaps they should be considered together.) Finally, it can be used to reason about the properties of the program itself.

In all three of these applications of Logic, the classical ideas must be supplemented by other insights, mathematical or otherwise. Traditionally, Logic is concerned with statements about broad types of reasoning - Godel's Theorems being the most famous - not about how it can be used to achieve certain purposes. There is a field called **computational logic**, which does address some of the issues necessary for such applications. I believe that the **algebraic** approach to logic - via Boolean and Heyting algebras, Categories and Toposes, sheaves, ... - has much to offer for any broad program of application, especially if its diffuse structure can be focussed in more concrete ways.

REMARKS ON AI AND SHEAVES

1. Introduction

A fundamental problem for AI researchers is to understand the nature and role of **time** and **dynamics** in Computer and AI systems, and find a substitute for the theory of differential equations and dynamical systems in traditional science and engineering.

The theory of Discrete Event Systems and Processes- originated in the Control context by Ramadge and Wonham [26, 8, 18] and in the CS framework by Milner, Hoare and Hennesey [14, 16] - provides a beginning towards such a methodology and mathematical framework. Following general principles suggested by Dana Scott [28-32], we might look to algebraic and topological ideas (partial orderings, lattices, categories, sheaves, Scott topologies, fixed points , ...) in order to provide a foundational framework that is mathematically capable of handling the wide variety of dynamical situations encountered in the hybrid Control/CS/ AI systems.

The first step in such a program is to try to describe as wide a spectrum as possible of dynamical systems and discrete-event systems in a common Category-theoretic framework. This will involve the theory of mathematical structures of **sheaf** and **category theory,** originating in differential and algebraic geometry, and carried over into Logic [5, 8, 10, 11, 12, 13, 20, 21, 22, 23, 24, 25, 27]. These tools have been used in Computer Science for the study of Semantics, but not yet for the sort of applications envisaged here : therefore, extensive preparatory development and exposition will be required. However, I believe that such mathematical tools will be useful for attainment of the goal of providing an integrated theory of Discrete and Continuous Event Dynamical Systems.

Here are some results that might follow from this research:

1) A unified (and better formulated) mathematical framework for the nascent hybrid computer science/ control discipline of **discrete event systems and processes.** The setting for this class of problem has so far been pragmatically dependent on its technological or computer science origin. The point of the 'systems' method is to develop a unified mathematical framework. This, after all, has been the source of much of the power of **automatic control theory** in the past forty years: A general mathematical theory of feedback, regulation, estimation, etc. was isolated from its origins in circuit and servomechanism theory, studied mathematically, and then applied much more widely and systematically. Our challenge now is to do the same job for the even broader class of phenomena engineers must deal with as a consequence of the diffusion of computer technology and the development of Computer Science.

2). Mathematical tools are needed that are capable of handling discrete' and 'continuous' phenomena in a more unified way. In pure mathematics, this is one of the main features of the geometric theory of Categories and Sheaves, as it evolved from the work of Ehresmann and Grothendieck in the 1950's. The triumph of this point of view was the proof by Deligne of the Weil Conjectures in number theory, which involved generalizing geometric technique developed for 'continuous' mathematics to the 'discrete' situation. This Unification is especially important for the sort of Autonomous Control engineering problems involved with advanced aircraft, robotics, etc, where the need will be to integrate the higher level 'discrete'; supervisory control with the lower-level, 'continous' system control associated with differential equations, transfer functions, etc.

3). **Discrete-event control theory** is, by its very mathematical nature, tied to the **ordering** properties of the integers or real numbers , rather than the analytical properties associated with modelling via differential equations. **Category theory** is that branch of algebra which is best adapted to the study of partial- orderings and its associated mappings. Dana Scott first used Lattice Theory in Compute Science in the development of the theory of Denotational Seman

The further evolution of this work has been in Categorical directions. I believe that the field of Discrete Event Systems will take a similiar evolution, for which the research I am proposing here will help lay the foundations.

4) **Artificial Intelligence** is the branch of Computer Science that also has great relevance to both the technological goals and the source of future research areas for Discrete Event Systems. After all, the Brain is the ultimate example of the DES! AI has its roots in difficult but ill-defined problems in logic, psychology and planning, and has consequently not had an extensive mathematical development and evolution since its beginnings in the 1950's. Applying and developing AI in the more tightly-conditioned environment of the engineer will require a more precise mathematical development: Category Theory is the appropriate mathematical language and tool (and it is both!) to systematize AI, and to integrate its insights into the more traditional differential-equations based methodologies of the engineering and physical science disciplines. For example, **Qualitative Physics** is the so-far primitive AI attempt to develop methods of reasoning about physical systems that might replace the traditional differential-equations based methods. This is the subject that is a prime candidate for development of mathematical methods using Category theory. Certainly, at the minimum, it will be useful to the CS/AI enterprise as a whole to diffuse knowledge of the algebraic Category methods from the small sub-culture that now is familiar with it to the CS/AI world at large. Further, the very 'distributed' nature of AI systems is captured mathematically very well by the notion of a Category, which unifies the theory of mathematical structures (ordered sets, groups, fiber bundles, ...) which have already proved to be useful in the physical and engineering sciences.

5) A better understanding of the process of describing a physical situation via a computer program, and the whole programming/compution process.

2. PRE-SHEAVES ON PARTIALLY-ORDERED SETS.

In the physical and engineering world, we are accustomed to regarding 'time' (whether 'discrete' or 'continuous') as a Lie group, with its combined analytic and algebraic structure. This structure is used as a mathematical foundation to construct differential and difference equations, dynamical systems, feedback control systems, etc. The Computer Science discipline of Artificial Intelligence has sought alternate, Computer Science-based methods replacing these traditional physical and engineering concepts associated with what mathematicians call the theory of Dynamical Systems.

Mathematically, physics and engineering is based on the **order** and **continuity** structure of the real numbers or integers, and the resulting mathematical theory of **dynamical systems.** We ask: How can dynamical systems be described in an algebraic way using these structures? **Category Theory** provides us with with a general mathematical methodology to do this. Exploring the mathematical consequences of these generalizations often leads to non-trivial unification, generalizations and insights.

For example, Causality, considered as a partial ordering, defines an example of the algebraic structure called a Category. We then might consider **functors** to other important Categories, topologies compatible with this partial ordering, pre-sheaves associated with such topologies, fixed points of continuous functors, etc. From the mathematical point of view Discrete Event Systems involves the combination of such a partial-ordered structure with the mathematical Language required to describe the Events. In terms of Category theory, it is a theory of **sheaves** over partially ordered sets.

I will sketch some elementary things that can be done with these ideas. Let us start withthe definition of a **pre-sheaf** on a partially-ordered set.

Consider a set T, with a partial ordering between its elements:

$$\{t \leq t'\} \qquad\qquad (2.1)$$

As the notation indicates, the reader should think of T as something like the real line or half-line, or the corresponding 'discrete' things, but other possibilities may be useful, for example in the theory of parallel and concurrent processes.

This algebraic structure is an example of a **Category** [4, 5, 9, 10, 11, 12, 13, 20, 21, 22, 23, 24, 25, 27]. The 'objects' are the points of T, and the 'arrows' are the ordered pairs (t, t') that satisfy 2.1. Another important Category is **SET**: The 'objects' are sets in the sense of 'naive' set theory (i.e. ignoring distinctions between 'sets', 'classes' , which logicians must employ in order to avoid the infamous logical and linguistic paradoxes) and the 'arrows' are maps between sets in the usual sense.

Definition.   **Pre-sheaves based on the category $\{T, \leq\}$ are** contravariant functors from $\{T, \leq\}$ to the category **SET**.

Let us make this more explicit.

Definition. A **pre-sheaf** on T assigns to each $t \in T$ a set $S(t)$ and to each pair $(t, t') \in T \times T$ such that $t \leq t'$ a map:

$$\phi(t, t'): S(t') \longrightarrow S(t). \tag{2.2}$$

The following version of the **associative law** is to be satisfied:

For each triple $(t, t', t'') \in T \times T \times T$ such that

$$t \leq t' \text{ and } t' \leq t'', \tag{2.3}$$

we have:

$$\phi(t, t'') = \phi(t, t')\phi(t', t''). \tag{2.4}$$

In addition, the following **identity law** is satisfied:

$$\phi(t, t) = \{\text{identity map}\} \text{ for each } t \in T. \tag{2.5}$$

As a Model, let us now think of a dynamical system in the usual sense, but interpreted in pre-sheaf terms.

3. THE PRE-SHEAF ASSOCIATED WITH A CONTINUOUS-TIME DYNAMICAL SYSTEM.

Consider a continuous-time dynamical system defined via a smooth ordinary differential equation of the following form:

$$dx/dt = f(x), \ x \ \varepsilon \ R^n, \tag{3.1}$$
$$t \ \varepsilon \ T \ = \ \text{non-negative real numbers}$$

Assign to the ODE 3.1 the following pre-sheaf over T:

To each $t \ \varepsilon \ T$, we assign the set $S(t)$ of all solution curves of 3.1 defined over the time - interval $[0, t]$. If $t \leq t'$, let

$$\phi(t, \ t'): \ S(t') \ \text{-->} \ S(t) \tag{3.2}$$

be the map which restricts the solution curves over $[0, t']$
to the interval $[0, t]$.

Re-defining the differential equation 3.1 as an integral equation, in the standard way, we can describe the situation in terms of the standard Scott theory [24, 28-32] of definition of Data Structures in terms of fixed points of continuouis mappings. (This is also related to what computer scientists call Denotational Semantics). Here is how this is set up in terms of Sheaf Theory:

Consider the pre-sheaf S over T which assigns to each $t \ \varepsilon \ T$ the set of continuous maps $C([0, t], R^n)$ from the interval $[0, t]$ to R^n. Make $C([0, t], R^n)$ into a topological space, using the sup-norm. This, together with an order-compatible topology on T, gives a topology for S. To each point $x_0 \ \varepsilon \ R^n$ we can assign a mapping:

$$P(x_0): S \longrightarrow S \qquad\qquad (3.3)$$

whose fixed points are the solutions $t \longrightarrow x(t)$ of 3.1 with the initial conditions:

$$x(0) = x_0 \qquad\qquad (3.4)$$

Continuity of $P(x_0)$ depends on a Lipschitz condition for $f(\ ,\ )$, in the usual Picard way.

Generalizing this procedure - defining topologies for pre-sheaves with an associated category of continuois maps - provides an interesting class of mathematical problems of a mixed toplogical/analytic nature which I plan to pursue further, perhaps using such tools of Global Analysis as the Atiyah-Singer Index Theorem. Notice that the standard dynamical- system- with-control, e.g. of the form:

$$dx/dt = f(x,\ u)$$
$$\qquad\qquad (3.5)$$
$$x \in R^n,\ u \in R^m$$

can be put into such a form, with:

$$S(t) = \{\text{maps of } [0,\ t] \text{ into } R^n \times R^m\} \qquad\qquad (3.6)$$

The Discrete Event Systems of Ramadge and Wonham [26, 18] can also be put into this framework, with T a general partially ordered set, a category C consisting of sublanguages of a given language, and the pre-sheaf S a contravariant functor from T to C. Developing these analogies within the framework of the nathematical theory of Categories and Sheaves offers exciting mathematical possibilities!

4. TRAJECTORIES OF A PRE-SHEAF AND 'TRACES' OF DISCRETE-EVENT SYSTEMS AND COMPUTER SCIENCE PROCESSES

Let us now turn to the development of a general theory of a 'discrete event system' or a' computer science process'. Let T be a set with a fixed partial-order relation $\leq$. Let $\{S(t), \phi(t, t'): S(t') \to S(t)\}$ be a sub-category of **Set**, the image of a contravariant functor from the category $\{T, \leq\}$ to **SET**. Let:

$$S = \{(s, t): s \ \varepsilon \ S(t), t \ \varepsilon \ T\}$$

The Cartesian projection map $(s, t) \to t$ defines a map: $S \to T$. It might be thought of geometrically as defining a **fiber space**.

Definition. Let T' be a subset of T. A map cross -section map $T' \to S, t \to (t, \gamma(t))$, with $\gamma(t) \ \varepsilon \ S(t)$, is called a **trajectory** of the **processes** associated to the Category $\{S(t), \phi(t, t'): S(t') \to S(t)\}$ if it satisfies the following relation:

$$\phi(t, t')(\gamma(t') = \gamma(t)$$

for all $(t, t') \ \varepsilon \ T' \ x \ T'$ such that:

$$t \leq t'.$$

Remark. In the computer science literature, what I have called 'trajectories' (following [15]) are called **traces**. My motivation for this change in terminology is precisely to suggest the geometric and dynamical-system nature of these objects.

The set of all such trajectories is denoted by $\Gamma(T')$. The assignment $T' \to \Gamma(T')$ defines another presheaf, based on the partially-ordered set of subsets of T. A **sub-presheaf** of the presheaf $\{\Gamma(T')\}$ is called a **process**.

It would now be appropriate to define functorial ways of specifying processes, analogous to the way 'algebraic varieties' are defined in the category-theoretic approach to algebraic geoemetry [25].   One Computer Science approach is to define the appropriate notion of **recursiveness**. (Note that this a major theme of [18]) Following Scott's ideas, this can be done by defining appropriate topologies on the presheaf {Γ(T')}, thereby defining the notion of **continuity** for functors between presheaves, and then defining the **recursive processes** as the fixed points of continuous functors.   This method has already been used in specical cases in various CS contexts [16, 14, 18]   Another method - also proposed in [18] - is to classify processes in terms of the 'complexity' of the 'languages' generated by the trajectories of the process.   In this context, the transition maps $\phi(t, t')$ defining the category may be   the analogue of the **production rules** of the language theorist and computer scientist.

5. POSSIBLE APPLICATIONS TO 'QUALITATIVE PHYSICS'

The category-theoretic approach to the description of dynamical systems   suggests an approach to the branch of AI called **qualitative physics** [19].   Equation 3.1 is a good 'model' of a physical system if "x" denotes an element of the **phase space** of the system.   For example, in **classical mechanics**, "x" might be the ordered pair (q, p) of a position vector q and a momentum vector p.   In **quantum mechanics**, "x" might be an element of a Hilbert space, etc.   The lingustic notions of 'position', 'velocity' 'momentum', etc. may now be thought of as **functors** from the category of subsets of phase space (or certain sub-categories defined by the knematics and dynamics) to certain **linguistic** and **logical** categories and sheaves. This can be made precise in what logicians call Model Theory [6].

In the usual formulations of physics, these functors are only used at the 'kinematic' level, to set up the dynamical equations of motion (via Lagrange's or Hamilton's equations). The pre-sheaf consisting of dynamical trajectories   is constructed as described in Section 3.

As I read it [19], 'qualitative physics' is an attempt to alternately define a pre-sheaf - which depends only on specified parts of the the 'linguistic' and 'logical sheaves' (such as 'something is increasing') - that in some sense **approximates** the 'exact' pre-sheaf of trajectories.

A precise formulation along these lines might serve to sharpen the scientific and mathematical structure of this discipline, as well as provide inspiration for the construction of useful computer algorithms. As a research beginning, I will start from the Model Theory [6] point of view: Replace the Language describing dynamical systems from the differential equations point of view with some sort of precise language reflecting the Qualitative Physics point of view, and then look for Models. Here too I suspect that the theory of Categories and Sheaves is the 'right' mathematical framework.

BIBLIOGRAPHY

1. H. Abelson and G. J. Sussman, Structure and Interpretation of Computer Programs, MIT Press, 1985.

2. A. Aho, J. Hopcroft and J. Ullman, Data Structures and Algorithms, Addison-Wesley, 1983.

3. S. Alagic and M. A. Arbib, The Design of Well-Constructed and Correct Programs, Springer-Verlag, New York, 1978

4. M. Arbib and E. Manes, Arrows, Structures and Functors, Academic Press, 1975.

5. M. Barr and C. Wells, Toposes, Triples and Theories, Springer-Verlag, New York, 1985

6. J. Bell and M. Machover, A Course in Mathem,atical Logic, North-Holland, Amsterdam, 1977.

7. W. Bibel and Ph. Jorrand, Fundamentals of Artificial Intelligence, Springer-Verlag, 1986.

8. Y. Brave and M. Heymann, Formulation and control of a class of real-time discrete-event processes, preprint, Ames Research Center, NASA, 1988

9. C. Ehresmann, Categories et Structures, Dunod, Paris, 1965

10. S. Eilenberg and S. MacLane, General theory of natural equivalences, Trans. Amer. Math. Soc., 58, 1945, 231-294.

11. S. Eilenberg and C. Elgot, Recursiveness, Academic Press, 1970

12. R. Goldblatt, Topoi, 2nd ed., N. Holland-Elsevier, 1984

13. W. S. Hatcher, The Logical Foundations of Mathematics, Pergamon Press, 1982

14. M. Hennessy, Algebraic Theory of Processes, MIT Press, 1988

15. R. Hermann, A 'geometric' view of the dynamics of the trajectories of computer programs, submitted to Acta Applicanda Mathematica

16. C. A. R. Hoare, Communicating Sequential Processes, Prentice Hall, 1985

17. J. Hopcroft and J. Ullman, Introduction to Automata Theory, Languages and Computation, Addison-Wesley, 1979

18. K. Inan and P. Varaiya, Finitely recursive process models for discrete event systems, IEEE Trans Aut. Con.,33, 626-638

19. J. de Kleer, Qualitative Physics, in Encyclopedia of Artificial Intelligence, J. Wiley, 1987.

20. J. Lambek and P. J. Scott, Introduction to Higher-Order Categorical Logic, Cambridge University Press, 1986.

21. S. MacLane, Categories for the Working Mathematician, Springer-Verlag, 1976

22. M. Makkai and G. Reyes, First Order Categorical Logic, Lecture Notes in Math. No. 611, Springer-Verlag, 1977

23. E. Manes, Algebraic Theories, Springer-Verlag, 1976.

24. E. G. Manes and M. A. Arbib, Algebraic Approaches to Program Semantics, Springer-Verlag, New York, 1986

25. D. Mumford, The Red Book of Varieties and Schemes, Springer-Verlag, 1988

26 . P. Ramadge and M. Wonham, Supervisory control of a class of discrete event processes, SIAM J. Control Opt., 25, pp. 206-230, 1987

27. G. Reyes, From Sheaves to Logic, in *Studies in Algebraic Logic*, ed. by A. Diagneault, MAA Studies in Mathematics, Vol. 9, 1974

28. D. S. Scott, Some ordered sets in computer science, in "Ordered Sets", I. Rival, Ed., Reidel, Dordrecht, 1982, pp. 677-718.

29. D. S. Scott, Data types as lattices, SIAM J. Comp., 5, 522-587, 1976

30. D. S. Scott, Logic and programming languages, Comm. ACM, 20, 634-641, 1977

31. D. S. Scott, Lectures on a mathematical theory of computation, in "Theoretical Foundations of Programming Methodology", M. Broy and G. Schmidt, Eds., Reidel, 1982.

32. J. E. Stoy, Denotational Semantics: The Scott-Strachey Approach to Programming Languages, MIT Press, 1977

MATH SCI PRESS
53 Jordan Road, Brookline, MA 02146 (USA)
(617) 738-0307

INTERDISCIPLINARY MATHEMATICS

by Robert Hermann

1. GENERAL ALGEBRAIC IDEAS. 1973.

2. LINEAR AND TENSOR ALGEBRA. 1973.

3. ALGEBRAIC TOPICS IN SYSTEM THEORY. 1973.

4. ENERGY MOMENTUM TENSORS. 1973.

5. TOPICS IN GENERAL RELATIVITY. 1973.

6. TOPICS IN THE MATHEMATICS OF QUANTUM MECHANICS, 1973.

7. SPINORS, CLIFFORD AND CAYLEY ALGEBRAS. 1974.

8. LINEAR SYSTEMS AND INTRODUCTORY ALGEBRAIC GEOMETRY. 1974.

9. GEOMETRIC STRUCTURE OF SYSTEMS-CONTROL THEORY AND PHYSICS, 1974.

10. GAUGE FIELDS AND CARTAN-EHRESMANN CONNECTIONS. 1975.

11. GEOMETRIC STRUCTURE OF SYSTEMS-CONTROL THEORY AND PHYSICS, 1976.

12. GEOMETRIC THEORY OF NON-LINEAR DIFFERENTIAL EQUATIONS, BAEKLUND
TRANSFORMATIONS AND SOLITONS, part A,   1976.

13. ALGEBRO-GEOMETRIC AND LIE-THEORETIC TECHNIQUES IN SYSTEM THEORY,
with C. Martin, 1977

14. GEOMETRIC THEORY OF NON-LINEAR DIFFERENTIAL EQUATIONS, BAEKLUND
TRANSFORMATIONS AND SOLITONS, part B. 1976. 3

15. TODA LATTICES, COSYMPLECTIC MANIFOLDS, BAEKLUND TRANSFORMATIONS   AND
KINKS, part A. 1977.

16. QUANTUM AND FERMION DIFFERENTIAL GEOMETRY. 1977.

17. DIFFERENTIAL GEOMETRY AND THE CALCULUS OF VARIATIONS, 2nd EDITION.
1977.

18. TODA LATTICES, COSYMPLECTIC MANIFOLDS, BAEKLUND TRANSFORMATIONS   AND
KINKS, part B. 1977.

19. YANG-MILLS, KALUZA-KLEIN AND THE EINSTEIN PROGRAM. 1978.

20. CARTANIAN GEOMETRY, NONLINEAR WAVES, AND CONTROL THEORY, part A.   1980.

21. CARTANIAN GEOMETRY, NONLINEAR WAVES, AND CONTROL THEORY, part B.   1980.

22. TOPICS IN THE GEOMETRIC THEORY OF LINEAR SYSTEMS. 1984.

23. TOPICS IN THE GEOMETRIC THEORY OF INTEGRABLE SYSTEMS. 1984.

24. TOPICS IN PHYSICAL GEOMETRY. 1988.

25. GEOMETRIC COMPUTING SCIENCE: FIRST STEPS. 1991.

LIE GROUPS: HISTORY, FRONTIERS AND APPLICATIONS

1. SOPHUS LIE'S 1880 TRANSFORMATION GROUP PAPER. Translated by M. Ackerman. Appendices by R. Hermann. 1975.

2. RICCI AND LEVI-CIVITA'S TENSOR ANALYSIS PAPER. Translated, edited, and comments by R. Hermann. 1975.

3. SOPHUS LIE'S 1884 DIFFERENTIAL INVARIANTS PAPER. Translated by M. Ackerman. Appendices by R. Hermann. 1975.

4. SMOOTH COMPACTIFICATION OF LOCALLY SYMMETRIC VARIETIES, by A. Ash, D. Mumford, M. Rapoport and Y. Tai. 1975.

6. THE 1976 AMES RESEARCH CENTER (NASA) CONFERENCE ON GEOMETRIC NONLINEAR WAVE THEORY. R. Hermann, Ed. 1977. $25.

7. THE 1976 AMES RESEARCH CENTER (NASA) CONFERENCE ON GEOMETRIC CONTROL THEORY, R. Hermann and C. Martin, Eds. 1977.

8. HILBERT'S PAPERS ON INVARIANT THEORY. Translated by M. Ackerman, comments by R. Hermann. 1978.

9. DEVELOPMENT OF MATHEMATICS IN THE 19TH CENTURY, by Felix Klein. Translated by M. Ackerman. Appendix: "Kleinian Mathematics from an Advanced Standpoint", by R. Hermann. 1979.

10. QUANTUM STATISTICAL MECHANICS AND LIE GROUP HARMONIC ANALYSIS, by N. Hurt and R. Hermann. 1980. .

11. FIRST WORKSHOP ON GRAND UNIFICATION, P. Frampton, S. Glashow, A. Yildiz, Eds. 1980.

12. INVERSE SCATTERING PAPERS: 1955-1963, by Irvin Kay and Harry E. Moses. 1982

13. GEOMETRY OF RIEMANNIAN SPACES, by Elie Cartan. Translation by M. Ackerman. Appendices by R. Hermann. 1983.

14. THE GEOMETRIC THEORY OF ORDINARY DIFFERENTIAL EQUATIONS, by Georges Valiron. Translation by James Glazebrook. Comments by R. Hermann. 1984.

15. THE CLASSICAL DIFFERENTIAL GEOMETRY OF CURVES AND SURFACES, by Georges Valiron. 1986

SYSTEMS INFORMATION AND CONTROL

1. GEOMETRY AND IDENTIFICATION: PREOCEEDINGS OF APSM WORKSHOP ON SYSTEM GEOMETRY, SYSTEM IDENTIFICATION, AND PARAMETER IDENTIFICATION. P. Caines and R. Hermann, Eds. 197 pages. 1983.
2. BERKELEY-AMES WORKSHOP ON NONLINEAR PROBLEMS IN CONTROL AND FLUID MECHANICS, L. R. Hunt and C. F. Martin, Eds. 1984.